# HYDRAULIC SYSTEMS ANALYSIS
## An Introduction

# HYDRAULIC SYSTEMS ANALYSIS
## AN INTRODUCTION

JOHN STRINGER
*Senior Lecturer,*
*Department of Mechanical Engineering,*
*University of Sheffield, United Kingdom*

**A HALSTED PRESS BOOK**

## JOHN WILEY & SONS
New York

First published in the U.K. in 1976 by
The Macmillan Press Ltd

Published in the U.S.A. by Halsted Press,
a division of John Wiley & Sons, Inc., New York

Printed in Great Britain

**Library of Congress Cataloging in Publication Data**

Stringer, John.
   Hydraulic systems analysis.

   Bibliography: p.
   Includes index.
   1. Hydraulic control.   2. System analysis.  I. Ti-
tle.
TJ843.S78      629.8′04′2       75–25885
ISBN  0–470–83377–7

# Preface

Engineering design calculations are usually based on approximations and simplifying assumptions rather than being exact and precise so as to ensure that results are obtained quickly and economically. Prerequisites of suitable calculation procedures are that they give results of sufficient accuracy for their purpose but are in no danger of being misleading. In the case of hydraulic systems and servomechanisms, one such technique for carrying out design calculations is based on the method of linear analysis. With this method the characteristics of the various components of any device are represented in the form of linear equations which are then taken together in order to predict how a completed system will behave. The actual behaviour of the practical device may be somewhat different from that predicted but the differences should be only in detail and not in essence.

This book traces the constituent features involved in applying linear methods to hydraulic mechanisms. A step-by-step approach is adopted with no attempt made to convert the text into a reference book. Involved are certain aspects of the subjects of dynamics and of fluid mechanics combined with the basic concepts of linear control theory and a few electrical ideas (for coping with electrohydraulics).

The text concentrates on this so-called 'small perturbation' or 'small excursion' or linearised analysis whose use in hydraulics has been developed in the past by various research workers. One pioneer in this field, under whose aegis the author has now worked for some years, is Professor J. K. Royle. More advanced techniques of analysis have also H. E. Merritt evolved and one book *Hydraulic Control Systems* by H. E. Merritt, dealing with both linear and other methods, is particularly recommended for further reading.

The author wishes to thank Mr K. Morris and Mr D. Puttergill for their help with diagrams, Mr K. H. Sutherland for his useful suggestions and Mrs Ivy Ashton for doing the typing.

J. D. Stringer
Sheffield, 1976

# Contents

# Nomenclature

| | |
|---|---|
| $a$ | cross-sectional area of pipe (also a coefficient) |
| $A$ | net area of piston |
| $B_c$ | magnetic flux density |
| $c_r$ | radial clearance |
| $C$ | capacitance |
| $C_1, C_2$ | constants |
| $C_d$ | discharge coefficient |
| $C_h$ | specific heat |
| $d$ | diameter |
| $D$ | $\equiv \mathrm{d}/\mathrm{d}t$ |
| $e$ | electrical potential |
| $E$ | Young's modulus |
| $f$ | fluid friction factor |
| $F$ | force |
| $g$ | gravity acceleration |
| $g_0$ | force/mass conversion factor |
| $G$ | an acceleration |
| $h$ | a coefficient |
| $i$ | $= (-1)^{1/2}$ |
| $I$ | electrical current |
| $I_c$ | electrical control current |
| $J$ | moment of inertia |
| $k$ | a constant |
| $K$ | a constant |
| $K_q$ | valve flow coefficient |
| $K_c$ | valve pressure–flow coefficient |
| $L$ | leakage coefficient |
| $m$ | a mass |
| $M$ | a mass |
| $n$ | angular speed (rad/s) |
| $N$ | a magnitude |
| $N_c$ | number of turns |
| $p$ | complex operator |
| $P$ | pressure or pressure difference (with various suffixes) |
| $q$ | volume flowrate |
| $\dot{q}$ | rate of change of $q$ |
| $Q$ | heat transfer rate or quantity of heat |
| $r$ | radius |

| | |
|---|---|
| $R$ | electrical resistance |
| $(Re)$ | Reynolds' number |
| $s$ | complex operator |
| $t$ | time |
| $T$ | time constant |
| $u$ | valve underlap |
| $u_1, u_2$ | specific internal energies |
| $v$ | velocity |
| $v_s$ | sonic velocity |
| $V$ | volume |
| $W_x$ | external work |
| $x$ | a displacement |
| $y$ | a displacement |
| $z$ | height above some datum |
| | |
| $\mathcal{f}$ | viscous friction rate |
| $\mathcal{G}$ | a gear ratio |
| $\mathcal{J}$ | Joule's equivalent |
| $\ell$ | length |
| $\mathcal{L}$ | inductance |
| $\mathcal{M}$ | amplitude ratio |
| $\mathcal{r}$ | a ratio |
| $\mathcal{t}$ | temperature |
| | |
| $\alpha$ | a ratio |
| $\alpha_p$ | pump capacity |
| $\beta$ | bulk modulus |
| $\gamma$ | angular deflection |
| $\delta_m$ | motor capacity |
| $\Delta$ | a change or difference of |
| $\varepsilon$ | eccentricity |
| $\zeta$ | damping ratio or factor |
| $\theta$ | a quantity with various suffixes |
| | ($\theta_i$ input, $\theta_o$ output, $\theta$ error) |
| $\lambda$ | see equation C.2 |
| $\mu$ | dynamic viscosity |
| $v$ | kinematic viscosity (also Poisson's ratio) |
| $\rho$ | density |
| $\sigma$ | $= 1/\beta$ |
| $\tau$ | torque |
| $\phi$ | phase angle |
| $\psi$ | poppet half angle |
| $\omega$ | a frequency (rad/s) |
| $\Omega$ | a constant velocity or rate of change |

# 1 Fluid Flow Calculations

The reader who is familiar with the traditional subject of 'hydraulics' will know most of the things dealt with in this first chapter; the aim here is to summarise those aspects of the subject which are particularly relevant when dealing with power hydraulic systems.

A basic approach is used as far as possible and in particular the steady flow energy equation is invoked rather than the 'Bernoulli' equation to help deal with the temperature rise which occurs owing to losses in a system. The chapter contains comments about pressure waves even though detailed calculations involving them are rarely necessary. Throughout this (and other) chapters, attention is concentrated on oil, but the methods of calculation are equally applicable to the flame-resistant fluids which are sometimes used instead of oil in hydraulic systems.

## 1.1 Power

The power available from a stream of oil can be taken as $P \times q$, where $P$ is the pressure of the oil and $q$ the volume flowrate. For example, the power available from a system operating at a pressure of 200 bar ($200 \times 10^5$ N/m$^2$) with a flowrate of 0.25 l/s ($0.25 \times 10^{-3}$ m$^3$/s) is 5 kW ($5 \times 10^3$ N m/s) or in British units 2901 lb/in$^2$ pressure, 15.26 in$^3$/s flowrate, gives 44 269 in lb/s or 6.71 hp (that is 5 kW).

For oil hydraulic systems using positive displacement type pumps and actuators (sometimes termed hydrostatic systems), any additional power available because of the elevation or the velocity of the fluid will be relatively so small as to be negligible.

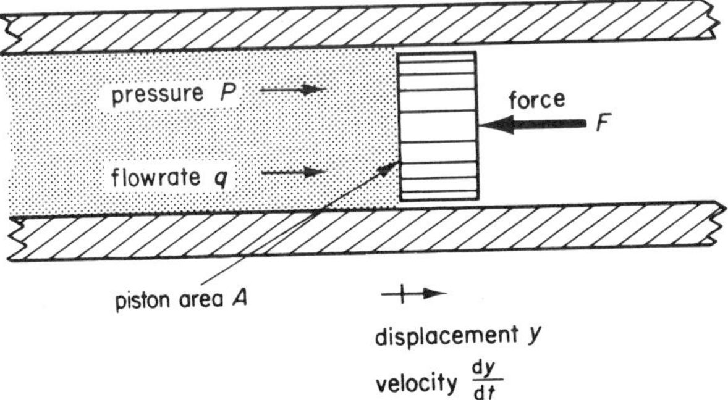


Figure 1.1 Piston in a long cylinder

To demonstrate that $P \times q$ is the available power, consider a long cylinder of cross-sectional area $A$ with a closely fitting piston as illustrated in figure 1.1. Fluid is flowing from left to right, its pressure is $P$ and the volume flowrate $q$. There is an external force $F$ resisting the piston motion.

To displace the piston to the right a distance $y$, the work done by the fluid against the resisting force $F$ would equal $Fy$. Power is the rate of doing work so the power being provided by the fluid would equal $F\,dy/dt$.

The maximum possible force $F$ which could be overcome by the fluid is $PA$ (that is with no friction). The maximum possible velocity $dy/dt$ at which the piston could be moved by the fluid is $q/A$ (that is with no leakage). Hence the maximum possible power which could be provided by the fluid is $PAq/A$ or $Pq$.

### 1.1.1   *Pumping*

Were the fluid stream being moved by the piston with fluid flowing in the opposite direction from that shown in figure 1.1, the power being provided by the force $F$ would be $F\,dy/dt$. In this case the minimum value of the force $F$ (with no friction) would be $PA$ and the minimum speed (with no leakage) would equal $q/A$; the least possible power for providing a flowrate $q$ at pressure $P$ would be $Pq$. For example, to pump 0.75 l/s to 70 bar requires at least 5250 W (that is 9.9 gal/min to 1015 lb/in$^2$ requires $9.9 \times 277 \times 1015$ in lb/min or 7.03 hp (5250 W), noting that one imperial gallon $\approx 277$ in$^3$). The actual power needed by a pump is always greater than $Pq$ owing to friction and leakage.

### 1.1.2   *Pressures*

A pump increases the pressure of a fluid stream. If the outlet pressure is $P_1$ and the inlet pressure $P_2$, then the minimum power required for a flowrate $q$ will be $(P_1 - P_2)q$.

With a motor or hydraulic piston, if the fluid enters at $P_1$ and leaves at $P_2$, then the maximum power obtainable will be $(P_1 - P_2)q$.

For a pump the outlet pressure is normally quoted as a 'gauge' pressure (that is pressure above the ambient or atmospheric pressure $P_{at}$) so that, when pumping from a tank at or near atmospheric pressure, the quoted outlet or delivery pressure is already equal to $P_1 - P_2$, where $P_1$ and $P_2$ are 'absolute' pressures and $P_2 \approx P_{at}$.

For motors or pistons, the lower pressure $P_2$ (in this case the outlet) is often greater than atmospheric pressure and has to be considered. For example, the maximum power available from a motor supplied with 0.1 l/s with inlet pressure 70 bar and outlet pressure 40 bar (both may be quoted as 'gauge' pressures) will equal $(70 - 40)10^5 \times 0.1 \times 10^{-3}$ or 300 W. In

British units 6.1 in³/s flowrate, 1015 lb/in² inlet and 580 lb/in² outlet gives 2650 in lb/s or 0.402 hp (that is 300 W).

For high-pressure systems, differences between gauge pressures and absolute pressures are often negligibly small. In cases of doubt absolute pressures should be used ($P_{at} \approx 15$ lb/in² $\approx 1$ bar).


## 1.2  The Steady Flow Equation

The power available from a fluid supply (equalling $Pq$) will usually be converted to mechanical form in a piston or motor. However, it may also be converted to kinetic energy as in an orifice; or it may be 'degraded' to appear as increased internal energy of the fluid—this would be sensed as an increased temperature. Calculations in these and other cases are more convenient on an energy basis (unit mass basis) than on a power basis (unit time) and a conversion simply involves dividing power by *mass* flowrate. The previously calculated power $Pq$ of a fluid stream becomes the 'pressure energy' of a unit mass of the fluid, namely $Pq/\rho q$ or $P/\rho$, where $\rho$ is the density of the fluid.

The kinetic energy of a unit mass of fluid moving at constant velocity $v$ is $v^2/2g_0$ (where $g_0$ is the force mass conversion factor for the particular units employed).

The internal energy of a unit mass of fluid is designated '$u$' and a change of such specific internal energy is designated $u_2 - u_1$. However, in the case of liquids, any change of internal energy is directly proportional to the change of temperature $t$ (almost exactly) so a change of internal energy may be written $C_h(t_2 - t_1)$—a conversion factor $\mathcal{J}$ usually being needed for $C_h$ (specific heat).

For a fluid passing through some device at a steady rate, the steady flow energy equation is applicable. (This equation is a statement of the conservation of energy law.) Fluid enters a device at state 1 (density $\rho_1$, specific internal energy $u_1$, absolute pressure $P_1$, velocity $v_1$, at elevation above some datum $z_1$) and leaves at state 2 ($\rho_2$, $u_2$, $P_2$, $v_2$, $z_2$), and the steady flow energy equation may be written

$$Q - W_x = u_2 - u_1 + \frac{P_2}{\rho_2} - \frac{P_1}{\rho_1} + \frac{v_2{}^2 - v_1{}^2}{2g_0} + (z_2 - z_1)\frac{g}{g_0}$$

where $Q$ is the heat transfer from the device per unit mass of fluid and $W_x$ the related work transfer. $g_0$ is the conversion factor and $g$ the gravity acceleration.

For oil hydraulic systems, the steady flow energy equation can usually be written

$$\frac{P_1}{\rho} = W_x + \frac{P_2}{\rho} + \frac{v_2{}^2}{2g_0} + \mathcal{J}C_h(t_2 - t_1) \tag{1.1}$$

(equation 1.1 is on a unit mass basis—multiply by $\rho q$ for converting to unit time (power) basis). $P_1$, $P_2$ are the inlet and outlet pressures respectively; $t_1$, $t_2$ are the inlet and outlet temperatures respectively; $v_2$ is the outlet velocity (assumed negligible inlet velocity); $W_x$ is the work done per unit mass (lb or kg) of fluid; $\rho$ is the density assumed to remain constant (any change with temperature or pressure assumed negligible). Heat transfer and changes in elevation are assumed negligible. $g_0 = 1$ for SI units, 32.2 for ft lb s units and $32.2 \times 12$ for in lb s units.

### 1.2.1 *Useful Forms of the Equation*

Equation 1.1 can be simplified in three distinct ways when one mode of energy transfer predominates.

*Case a*   For mechanical energy conversion as in a ram or motor, the term $W_x$ predominates with kinetic and internal energy changes being negligible so that equation 1.1 could be written

$$W_x = \frac{P_1 - P_2}{\rho} \tag{1.1a}$$

but the more convenient form in this case is the unit time (power) basis already considered. (Power $= \rho q W_x = (P_1 - P_2)q$.)

*Case b*   For fluid being accelerated, as through an orifice for example, changes of internal energy are usually negligible and there is no work transfer so that equation 1.1 may be written

$$\frac{v_2^{\,2}}{2g_0} = \frac{P_1 - P_2}{\rho} \tag{1.1b}$$

*Case c*   For energy being completely 'degraded', by passing fluid through some restriction into a large tank, for example, any kinetic energy being completely dissipated, and with no work transfer involved we may write

$$\mathscr{J}C_h(t_1 - t_2) = \frac{P_1 - P_2}{\rho} \tag{1.1c}$$

### 1.2.2 *Orifices*

The energy equation in the form of equation 1.1b is used for orifice calculations and is often written (*with the $g_0$ factor omitted*) as

$$v = \left\{ \frac{2(P_1 - P_2)}{\rho} \right\}^{1/2}$$

where $P_1 - P_2$ is the pressure drop across the orifice.

1.2.2.1 *Velocity* For example, the oil velocity through an orifice with a pressure drop across it of 30 bar (435 lb/in$^2$) assuming the oil density is 800 kg/m$^3$ (0.029 lb/in$^3$, that is specific gravity 0.8) is given by

$$v = \left(\frac{2 \times 1 \times 30 \times 10^5}{800}\right)^{1/2} = 86.6 \text{ m/s (SI units with } g_0 = 1)$$

or

$$v = \left(\frac{2 \times 386 \times 435}{0.029}\right)^{1/2} = 3400 \text{ in/s (in lb s units with } g_0 = 386)$$

1.2.2.2 *Flowrate* Volume flowrates through orifices are usually based on velocities calculated in this way although actual velocities may be slightly lower (by say 2%, that is $C_v \approx 0.98$ usually approximated to 1). With a high Reynolds number (based on orifice size) which is usual in oil system orifices, the area of the jet in which the fluid is moving at velocity $v_2$ (at the vena contracta where, strictly, $P_2$ should be measured) is often taken as $\frac{5}{8}$ of the orifice actual area so the volume flowrate is given by

$$q = C_d A_0 v_2$$

where $C_d = \frac{5}{8}$

$$= C_d A_0 (P_1 - P_2)^{1/2} \left(\frac{2}{\rho}\right)^{1/2} \tag{1.2}$$

and with SI units, for pressures in bars, this reduces to

$$q \approx 10 A_0 (P_1 - P_2)^{1/2} \tag{1.2a}$$

(or with in lb s units $q \approx 100 A_0 (P_1 - P_2)^{1/2}$)—note that $P_2$ is usually measured well downstream of the orifice.

For example, the oil flowrate through a 5 mm (0.197 in) diameter orifice with a pressure drop across it of 30 bar (435 lb/in$^2$), assuming specific gravity 0.8, would be given by equation 1.2 as

$$q = \frac{5}{8} \frac{\pi}{4} \frac{5^2}{10^6} \left(\frac{2 \times 1 \times 30 \times 10^5}{800}\right)^{1/2} = 1.06 \times 10^{-3} \text{ m}^3/\text{s} = 1.06 \text{ l/s}$$

or in British units

$$q = \frac{5}{8} \frac{\pi}{4} (0.197)^2 \left(\frac{2 \times 386 \times 435}{0.029}\right)^{1/2} = 64.8 \text{ in}^3/\text{s}$$

Alternatively from equation 1.2a

$$q = 10 \times \frac{\pi}{4} \times \frac{5^2}{10^6} (30)^{1/2} = 1.08 \times 10^{-3} \text{ m}^3/\text{s}$$

or in lb s units $100 (\pi/4)(0.197)^2 (435)^{1/2} = 63.6 \text{ in}^3/\text{s}$.

### 1.2.3   *Oil Temperature Rise*

Oil pressures decrease along pipelines (and through pipe fittings, valves and other restrictions) because of fluid friction. Pressures may decrease to near atmospheric pressure through some restriction into a tank (where any kinetic energy is dissipated) or owing to leakage through clearance gaps. In all cases, the fluid temperature will rise and, assuming there is no heat transfer from the fluid, the temperature rise may be estimated using the energy equation in the form

$$\mathcal{J}C_h(t_2 - t_1) = \frac{P_1 - P_2}{\rho} \tag{1.1c}$$

where $t_1$, $P_1$ and $t_2$, $P_2$ are the temperatures and pressures before and after the restriction, $C_h$ is the specific heat of the fluid and $\mathcal{J}$ a conversion factor.

With many oils a drop in pressure of 16–18 bar causes a rise in temperature of 1 degC (130–150 lb/in$^2$ causing about 1 degF).

The value of Joule's equivalent $\mathcal{J}$ is unity if the specific heat is expressed in J/kg degC and, as a numerical example, the temperature rise incurred when oil of specific heat 2093 J/kg degC (that is 0.5 cal/g degC or 0.5 Btu/lb degF) is throttled through a relief valve from 69 bar (100 lb/in$^2$), assuming specific gravity 0.8, may be estimated as

$$2090(t_1 - t_2) = \frac{69 \times 10^5}{800}$$

$$t_1 - t_2 = 4.1 \text{ degC}$$

or in British units noting that $\mathcal{J}$ is 9339 in lb = 1 Btu

$$9339 \times 0.5(t_1 - t_2) = \frac{1000}{0.029}$$

$$t_1 - t_2 = 7.4 \text{ degF (that is 4.1 degC)}$$

Note that the drop in pressure is taken to be the total 69 bar, that is $P_1$, the system pressure is 69 bar gauge pressure and $P_2$ is zero gauge pressure.

## 1.3  Flowrates

The pressure drop across an orifice (within a metering valve for example) may represent a significant proportion of the system operating pressure. The pressure drop along connecting pipelines is of interest in the suction line to pumps and in small-bore (restrictor) tubes provided as dampers—across rams and motors for example. Another leakage flow of interest is that through annular gaps as between valve spools or pump pistons and their mating bores.

### 1.3.1 *Flow through Tubes*

Flow along a tube will be laminar (or streamline) if the nominal fluid velocity (that is the volume flowrate divided by the tube cross-sectional area) is below a certain value. This value depends on the internal pipe diameter $d$ and the kinematic viscosity $v$ of the fluid. The Reynolds number is

$$(Re) = \frac{vd}{v}$$

where $v$ is nominal velocity (conveniently in cm/s), $d$ is the (internal) pipe diameter (conveniently in cm) and $v$ is the kinematic viscosity (conveniently in 'stokes' (St)—it is usually accepted that $(Re)$ must be less than 2000 for laminar flow to occur.

As a numerical example, consider the Reynolds number for oil of viscosity 100 cSt (that is 1 St or $1 \times 10^{-4}$ m$^2$/s) flowing through a 25.4 mm inside diameter pipe with a nominal velocity of 4.5 m/s (450 cm/s).

Using cm notation

$$(Re) = \frac{450 \times 2.54}{1} = 1143$$

Using SI notation

$$(Re) = \frac{4.5 \times 25.4 \times 10^{-3}}{1 \times 10^{-4}} = 1143$$

Calculations may be carried out using in lb s units, when kinematic viscosity is converted from stokes by, in effect, converting cm$^2$ to in$^2$, that is $1/(2.54)^2 = 0.155$ and 1 stokes $= 0.155$ units with this system. For the above example, $v = 177$ in/s and $d = 1$ in giving

$$(Re) = \frac{177 \times 1}{0.155} = 1143$$

For connecting pipelines in oil systems nominal oil velocities are traditionally kept below about 4.5 m/s. For oil of viscosity 100 cSt, $(Re)$ will be less than 2000 at this velocity only if the pipe is smaller than about 44 mm that is 2000/45). For oil of viscosity 10 cSt at this velocity the diameter would have to be less than 4.4 mm. Hydraulic oil viscosities usually lie in the range 10–100 cSt.

Fully developed laminar flow of a fluid at a constant rate through a smooth straight pipe of constant diameter exhibits a paraboloid velocity profile. The average velocity of flow is directly proportional to the drop in pressure per unit length of the pipe and to the square of the pipe radius or diameter, and it is inversely proportional to the absolute viscosity of the fluid. Average velocity times pipe cross-sectional area gives volume flowrate. For a length of pipe $l$ with pressure drop from $P_1$ to $P_2$ the relations are

$$v_{av} = \frac{P_1 - P_2}{\ell} \frac{1}{8\mu} r^2$$

$$q = \frac{P_1 - P_2}{\ell} \frac{1}{8\mu} \pi r^4 \tag{1.3}$$

(where $r$ is the internal radius of the pipe).

### 1.3.2 *Capillary (Small-bore) Tubes*

Small-bore tubes are used as restrictors in oil systems. The Reynolds number is usually less than 2000 and equation 1.3 may be used to predict the flowrates through them. For example, if the pressure drop along a 0.1 m (3.94 in) length of 1.5 mm (0.06 in) internal diameter tube carrying oil of viscosity 100 cSt and of specific gravity 0.8 is 35 bar (508 lb/in$^2$) then the flowrate may be estimated as

$$q = \frac{35 \times 10^5}{0.1} \frac{1}{8 \times 1 \times 10^{-4} \times 800} \pi \left(\frac{1.5 \times 10^{-3}}{2}\right)^4$$

$$= 5.4 \times 10^{-5} \text{ m}^3/\text{s (or 0.054 l/s)}$$

or in British units

$$q = \frac{508\pi(3 \times 10^{-2})^4}{3.94 \times 8 \times 80 \times 14.5 \times 10^{-8}}$$

$$= 3.48 \text{ in}^3/\text{s}$$

(noting that centistokes times specific gravity gives centipoise numerically and that 14.5 reynolds $= 10^6$ poises (P)).

### 1.3.3 *Connecting Pipelines*

With laminar flow in connecting pipelines, the pressure losses are usually small. For example, consider a 1 m (39.4 in) length of 25.4 mm (1 in) inside diameter pipe carrying 0.5 l/s (6.6 gal/min or 30.5 in$^3$/s) of oil with viscosity 50 cSt and specific gravity 0.87 for which ($Re$) would equal 500, using equation 1.3 rewritten as

$$P_1 - P_2 = \frac{q \times \ell \times 8 \times \mu}{\pi r^4}$$

$$= \frac{0.5 \times 10^{-3} \times 1 \times 8 \times 870 \times 0.5 \times 10^{-4}}{\pi(12.7 \times 10^{-3})^4}$$

$$= 2130 \text{ N/m}^2$$
$$= 0.0213 \text{ bar}$$

or in British units

$$P_1 - P_2 = \frac{30.5 \times 39.4 \times 8 \times 0.000\,014\,5 \times 0.5 \times 0.87}{\pi \times 1/16}$$

$$= 0.31 \text{ lb/in}^2$$

### 1.3.4 Turbulent Flow

When $(Re)$ exceeds about 2000, an empirical relation is used for calculating pressure drop in pipelines involving the 'friction factor' $f$. The value of $f$ may be taken as 0.01 or alternatively as $0.08/(Re)^{1/4}$. The empirical relation is

$$\frac{P_1 - P_2}{\rho} = \frac{4f\,\ell v^2}{2g_0 d} \tag{1.4}$$

For example, a 4 m (13.1 ft) length of 25.4 mm internal diameter pipeline carrying 2.5 l/s (153 in³/s) of oil with viscosity 35 cSt and having specific gravity 0.87 would have $(Re) = 3580$ (giving $f = 0.0103$) and

$$P_1 - P_2 = \frac{4 \times 0.0103 \times 4 \times 870}{2 \times 1 \times 25.4 \times 10^{-3}} \left\{ \frac{2.5 \times 10^{-3}}{\pi(12.7 \times 10^{-3})^2} \right\}^2$$

$$= 68\,700 \text{ N/m}$$

$$= 0.687 \text{ bar}$$

or in British units

$$P_1 - P_2 = \frac{4 \times 0.0103 \times 13.1 \times 12 \times 0.0314}{2 \times 386 \times 1} \left( \frac{153 \times 4}{\pi} \right)^2$$

$$= 10 \text{ lb/in}^2$$

(density 0.0314 lb/in³ and $g_0 = 386$).

### 1.3.5 Approximations Used in Calculation

Calculations based on equation 1.3 refer to fully developed laminar flow. They are, however, commonly applied for unsteady flow conditions. They are also commonly used for discrete pipe lengths although a certain transition length is needed for laminar flow to develop even under steady conditions (this length may be estimated as $0.0575d(Re)$). Later in this book certain calculations will be given based on the assumption that a flowrate $q$ is directly proportional to a pressure difference $P$ using equations of the form $q = LP$. The value of the constant $L$ will be taken as that calculated from the conventional (steady flow) relationship for viscous or laminar flow and based on the normally accepted parabolic velocity profile for liquid flowing at a steady rate. This will be done despite the fact that the flow of oil to and from an actuator in a hydraulic system is rarely steady or continuous for very long. To control the motion of whatever load is being moved hydraulically involves modulating and possibly reversing the flow. A changing flowrate produces varying velocity profiles but the effects of such variations in the profile will be neglected in the interests of simplicity.

### 1.3.6 *Annular Passages*

The radial clearance $c_r$ between a piston and its cylinder or a valve spool and its bore in an oil system will be in the region of 5 $\mu$m (0.0002 in) and ($Re$) will always be low. Flow along the related annular passages may be treated as flow between two parallel plates of width $\pi d$ and separated by distance $c_r$ if we assume the assembly to be concentric.

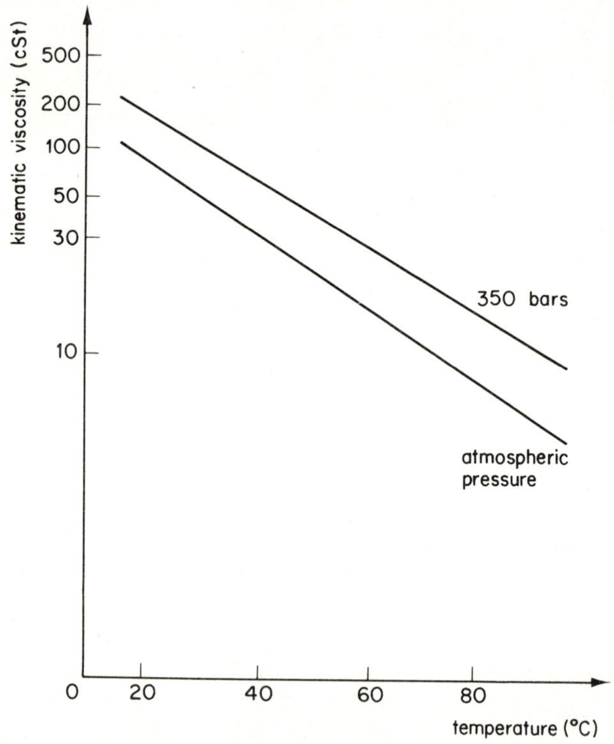



Figure 1.2 Viscosity versus temperature for a typical hydraulic oil (after *Technical Data on Shell Tellus Oil*, by permission of Shell International Petroleum Company Limited)

The relation between volume flowrate $q$ and pressure drop $P_1 - P_2$ for a contact length $\ell$ is in this case

$$q = \frac{P_1 - P_2}{\ell} \frac{\pi dc_r^3}{12\mu} \tag{1.5}$$

If the spool or piston lies eccentrically in its bore, however, the flowrate is greater and can attain $2\frac{1}{2}$ times the concentric value. The relation, for a relative eccentricity between the two axes of $\varepsilon$ is (Merritt, 1967)

$$q = \frac{P_1 - P_2}{\ell} \frac{\pi dc_r^3}{12\mu} \left\{ 1 + \frac{3}{2}\left(\frac{\varepsilon}{c_r}\right)^2 \right\}$$

### 1.3.7  *Viscosity Variations*

Viscosity-based calculations can only be as accurate as the value assigned to the viscosity. In practice a suitable value is difficult to select because oil viscosities vary significantly with temperature (and to a lesser extent with pressure—see figure 1.2, for example) and flowrates are often underestimated.

## 1.4  Compressibility

Transient flowrates in oil systems are often associated with oil compressibility (that is flow associated with pressure changes rather than with movement). To pressurise an already full container requires an extra volume of oil to be pumped in. Compressibility $\sigma$ is defined as the change in volume per unit volume for unit change in pressure. Bulk modulus $\beta$ is the reciprocal of compressibility. For many oils the bulk modulus $\beta_o$ can be taken as constant and equal to about $17 \times 10^8$ N/m$^2$ (say 247 000 lb/in$^2$) although the 'effective' bulk modulus $\beta$ for oil in a system or container will be less than this owing to dilations of the container and the possible presence of air in the oil.

With steady flow we need not be concerned with compressibility effects because actual density changes are small (for example less than $\frac{1}{2}\%$ for a change in pressure of 69 bar or 1000 lb/in$^2$).

Flowrates both transient and steady are usually expressed as volume rates $q$.

### 1.4.1  *Compressibility Flowrates*

To increase the pressure of the contents of a rigid vessel of volume $V$ from $P_1$ to $P_2$ requires an extra volume of oil to be pumped in equalling $(P_1 - P_2)V/\beta_o$. This volume of extra oil may be small but, if the pressure has to be increased rapidly, then the (transient) *rate of flow* may be very large and possibly exceed (for a short period) the capacity of the pump in a system. The *rate* of oil flow into a rigid container of volume $V$ whose pressure was increasing at a rate $dP/dt$ would equal $(dP/dt) \times V/\beta_o$.

As a numerical example, oil at 7 bar (102 lb/in$^2$) in a rigid vessel of volume 1 litre (61 in$^3$) is increased in pressure to 147 bar (2132 lb/in$^2$), and taking $\beta_o$ as $17 \times 10^8$ N/m$^2$ (246 600 lb/in$^2$), the volume which must be pumped into the container is $(147 - 7) \times 10^5 \times 1/(17 \times 10^8)$ or about 0.008 1 or in British units $(2132 - 102) \times 61/246\,600$ or $\frac{1}{2}$ in$^3$. Continuing the example, if this pressure has to rise in, say, 1/20 s (and this could represent a large proportion of the cycle time of some hydraulic machines), then the average flowrate during that short period would be 0.16 l/s (10 in$^3$/s).

Similar calculations relate to the reduction of pressure with flowrates from high-pressure oil volumes.

Transient flowrates associated with oil compressibility are likely to be significant compared with other flowrates in a system. Such flowrates are proportional to rates of change of pressure and may be expressed as

$$q_c = \frac{V}{\beta}\frac{\mathrm{d}P}{\mathrm{d}t} \tag{1.6}$$

(where $V$ is the volume of oil involved, $\beta$ is the effective bulk modulus and $\mathrm{d}P/\mathrm{d}t$ is the rate of change of pressure of the oil).

### 1.4.2    *Dilation of Containers*

An increase of oil pressure causes a dilation of the enclosing vessel, a pipe, for example, or the cylinder of a hydraulic ram. This dilation effectively reduces the bulk modulus of the fluid. For a cylindrical container of inside diameter $d_1$, outside diameter $d_2$, made of material with Young's modulus $E$ and Poisson's ratio $v$, the effective bulk modulus is given by

$$\frac{1}{\beta} = \frac{1}{\beta_o} + \frac{d_1 + d_2}{E(d_1 - d_2)}$$

for thin-walled cylinders (thickness less than $d/10$, say) or

$$\frac{1}{\beta} = \frac{1}{\beta_o} + \frac{2}{E}\left(\frac{d_2{}^2 + d_1{}^2}{d_2{}^2 - d_1{}^2} + v\right)$$

for thick-walled cylinders.

An example is the case of a copper tube for which $E = 8.274 \times 10^{10}$ N/m$^2$ ($12 \times 10^6$ lb/in$^2$) and $v = 0.33$ if the inside diameter is 12.7 mm (0.5 in) and the outside diameter is 15.9 mm (0.625 in) containing oil of bulk modulus $17 \times 10^8$ N/m$^2$ (246 600 lb/in$^2$). The ratio of the internal radius to wall thickness, 6.35/3.2, indicates 'thick wall' and the final term of the above equation equals $1.1734 \times 10^{-10}$ ($0.814 \times 10^{-6}$ British units) whilst for oil $1/\beta_o = 5.88 \times 10^{-10}$ ($4.06 \times 10^{-6}$ British units) which added give $7.05 \times 10^{-10}$ ($4.87 \times 10^{-6}$ British units) so the effective bulk modulus would be $14.2 \times 10^8$ (205 000 lb/in$^2$) representing a reduction (compared with oil alone) of about $16\frac{1}{2}\%$.

### 1.4.3 *Air Content*

A drastic reduction of effective bulk modulus occurs if free air is present in an oil system. It is generally considered that air in solution has no effect.

The effect can be illustrated by assigning a 'bulk modulus' to the air itself. For pressures varying about some mean value $P$, this value of 'bulk modulus' is in fact equal to $P$. Knowing the volume of air present per unit volume of oil and assuming a *perfectly rigid container*, the effective bulk modulus is estimated from

$$\frac{1}{\beta} = \frac{1}{\beta_o} + \frac{V_a}{V_o}\frac{1}{P}$$

For example with 0.1% of air by volume (1 part per thousand) and with pressures fluctuating around 17 bar (247 lb/in$^2$) the effective bulk modulus would be given by

$$\frac{1}{\beta} = \frac{1}{17 \times 10^8} + \frac{1}{1000}\frac{1}{17 \times 10^5} = \frac{1}{8.5 \times 10^8}$$

or the effective bulk modulus would be *halved* (the reduction being less at higher pressure). For British units

$$\frac{1}{\beta} = \frac{1}{246\,600} + \frac{1}{1000}\frac{1}{247} \approx \frac{1}{123\,300}$$

To justify assigning this 'bulk modulus' $P$, consider a volume $V_a$ of air dispersed (as bubbles) in a volume $V_o$ of oil. With expansions and contractions of the air occurring at constant temperature (the bulk oil temperature) then, if $PV = \text{constant}$ for the air, a decrease in air volume would be in

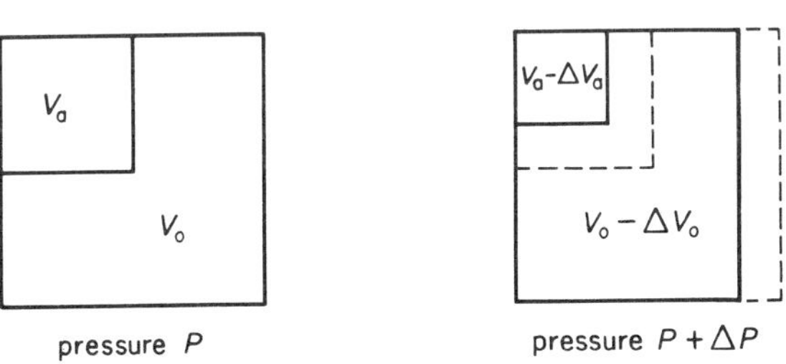


Figure 1.3 Oil with free air

exactly the same proportion as an increase in the pressure or

$$P\,\Delta V_a = -\,V_a\Delta P \text{ or } \Delta V_a = -\frac{V_a}{\beta_a}\Delta P$$

where $\beta_a = P$.

To justify the effective bulk modulus calculations, consider the air and oil separated (but at the same temperature) as shown in figure 1.3.

The effective bulk modulus is given by

$$\frac{1}{\beta}\Delta P = -\frac{\Delta V_a + \Delta V_o}{V_a + V_o}$$

$$= -\frac{V_a}{V_a + V_o}\frac{\Delta V_a}{V_a} - \frac{V_o}{V_a + V_o}\frac{\Delta V_o}{V_o}$$

$$= \frac{V_a}{V_a + V_o}\frac{\Delta P}{\beta_a} + \frac{V_o}{V_a + V_o}\frac{\Delta P}{\beta_o}$$

and, as $V_a$ is normally small (say 0.01% to 1% of $V_o$) and also $\beta_a = P$, we have

$$\frac{1}{\beta} \approx \frac{V_a}{V_o}\frac{1}{P} + \frac{1}{\beta_o}$$

Other methods are also used for estimating the effects of free air (see, for example, Smith *et al.* (1960)).

### 1.4.4  *Flexible Hoses*

A drastic reduction of effective bulk modulus occurs owing to the expansions of flexible hose. 'Bulk modulus' values for hose can be computed from manufacturers' data and may be less than $7 \times 10^8$ N/m$^2$ (100 000 lb/in$^2$).

### 1.4.5  *Surges*

The sudden closing of a valve to stop a flowing fluid causes a pressure surge. For the particular case of closing a valve at the outlet end of a pipeline, the rise in pressure may be predicted as

$$\Delta P = \frac{\rho}{g_0}v_s v$$

where $\rho$ is the fluid density, $v_s = \{(\beta/\rho) \times g_0\}^{1/2}$ and $v$ is the initial velocity (assumed constant) of all elements of the fluid. (The equation is modified to $\Delta P = 1.33\,\rho v_s v/g_0$ for fluid initially flowing along a circular pipe under laminar conditions, where $v$ is the *average* initial velocity of the fluid.)

For example, the pressure rise incurred by suddenly stopping oil of specific gravity 0.87 and bulk modulus $17 \times 10^8$ N/m² (246 600 lb/in²) moving in a perfectly rigid pipe at (constant) velocity 4.5 m/s (177 in/s) is obtained as follows:

$$v_s^2 = \frac{17 \times 10^8}{870}$$

$$v_s = 1398 \text{ m/s}$$

or in British units

$$v_s^2 = \frac{246\,600 \times 386}{0.0314}$$

$$v_s = 55\,060 \text{ in/s (4588 ft/s)}$$

Hence $v_s v \rho / g_0$ is

$$1398 \times 4.5 \times 870 = 5.47 \times 10^6 \text{ N/m}^2$$
$$= 54.7 \text{ bar}$$

or

$$\frac{55\,060 \times 177 \times 0.0314}{386} = 793 \text{ lb/in}^2$$

With the same oil in a flexible container (or with free air present—or both) such that the effective bulk modulus is $4 \times 10^8$ N/m² (58 000 lb/in²), the result is 26.5 bar (385 lb/in²) pressure rise. A pressure rise of 750 lb/in² or 51.7 bar for 15 ft/s or 4.57 m/s flow velocity is mentioned by Merritt (1967). Incidentally, water has a higher bulk modulus (about $21 \times 10^8$ N/m² or 300 000 lb/in²) and higher density (specific gravity $= 1$) giving the result for 4.5 m/s in a rigid container of about 65 bar (946 lb/in²).

This method of estimating the surge or rise in pressure due to the sudden closing of a valve is based on the kinetic energy of the fluid being wholly converted to strain energy. The strain energy of a mass of stationary fluid fluid at pressure $P_1$ occupying volume $V_0$ is equal to the work which could be done by the fluid as it expanded to zero pressure. This work is given by $\int_{P_1}^0 P \, dV$ or, with the assumed (constant bulk modulus) relation for the fluid, namely $\delta V = (V_0/\beta)\delta P$; this work equals $(V_0/\beta)(P_1^2/2)$. Similarly the increase in strain energy of a mass of fluid due to a pressure increase of $\Delta P$ is given by $(V_0/\beta)\{(\Delta P)^2/2\}$. In the case of a pipeline of length $\ell$ and cross-sectional area $a$, the term $V_0$ is equal to $\ell a$. The kinetic energy of a mass of fluid in such a pipeline assuming each element of fluid is moving at constant velocity $v$ is $\frac{1}{2}(mv^2/g_0)$ or $\frac{1}{2}(\rho/g_0)\ell a v^2$. A sudden closure causes the fluid to stop and equating this kinetic energy to the strain energy gives the pressure rise

$$\frac{1}{2}\frac{\rho}{g_0}\ell a v^2 = \frac{1}{2}\frac{\ell a}{\beta}(\Delta P)^2$$

or

$$\Delta P = v v_{\mathrm{s}}\frac{\rho}{g_0}$$

where

$$v_{\mathrm{s}} = \left(\frac{\beta}{\rho}\times g_0\right)^{1/2}$$

### 1.4.6 *Pressure Waves*

A disturbance such as a pressure surge travels through a fluid at a velocity which depends (almost entirely) on the effective bulk modulus of the fluid and its density. This 'sonic' velocity is

$$v_{\mathrm{s}} = \left(\frac{\beta}{\rho}\times g_0\right)^{1/2}$$

and a numerical example has already been quoted which gave about 1400 m/s (for a sample oil in a rigid container). Velocities in the range 900–1200 m/s (3000–4000 ft/s) may be expected in oil systems with steel pipelines and without unduly large air content.

The time for a disturbance to travel along an oil column of length $\ell$ (contained in a pipe, for example) is $\ell/v_{\mathrm{s}}$.

A pressure 'wave' would occur in such a column as the result of successive reflections at the ends; reflections in the positive sense at closed ends and in the negative sense at open ends.

The frequency of a pressure wave in a column of oil length $\ell$ will equal $v_{\mathrm{s}}/4\ell$, the 'organ pipe' frequency associated with one open end and one closed end. (Another frequency which may occur is $v_{\mathrm{s}}/2\ell$.)

For example, the frequency to be expected in a pipeline $\ell$ metres long for $v_{\mathrm{s}} = 1000$ m/s is $1000/4\ell$ Hz. For a length $\ell = 2.5$ m, this frequency is 100 Hz.

Pressure waves occur in many circumstances. Two are considered below.

(i) The pressure $P_{\mathrm{d}}$ induced at the closed end of a pipe when the pressure $P_{\mathrm{u}}$ at the other end is suddenly increased takes the form illustrated in figure 1.4.

(ii) The pressure induced at the face of a piston of area $a$ pushing a column of oil (which is also of area $a$ and has mass $m$) along inside a pipe—assuming the piston suddenly starts moving with constant acceleration $G$ and assuming the other end to be open—takes the form shown in figure 1.5. (Note that,

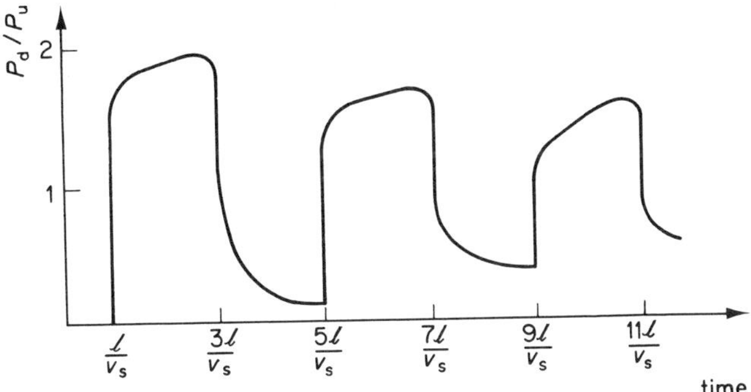

Figure 1.4  Pressure wave in a closed pipe

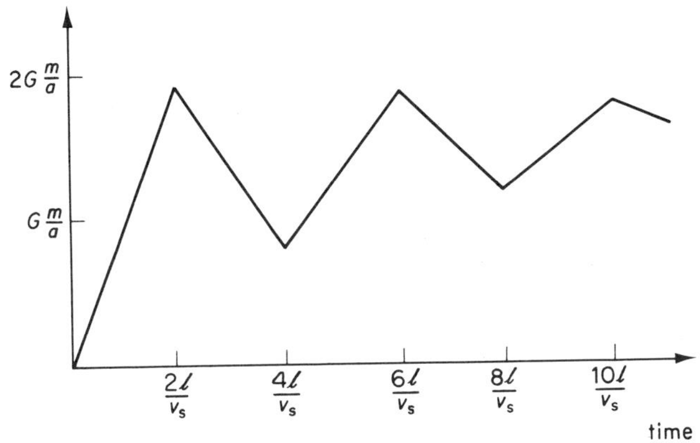

Figure 1.5  Pressure on piston face (after Leedham (1965))

without pressure waves, the pressure would equal $Gm/a$—or $1\tfrac{1}{3}GM/a$ for a fully developed laminar flow—plus that (increasing) pressure needed to overcome fluid friction in the pipe.)

Detailed studies of transient and pulsating flows have been made particularly by Helmholz (Crandall, 1926), and by workers at MIT and at Birmingham University (Brown and Nelson, 1965; Foster and Parker, 1964–5; see also Lorenz and Stringer, 1966).

## Problems

(Assume oil has specific gravity of 0.87 and bulk modulus of $17 \times 10^8$ N/m$^2$ or 246 600 lb/in$^2$.)

1 Estimate the power required to drive a pump drawing oil from a tank at atmospheric pressure if the flowrate is 0.8 l/s (48.8 in$^3$/s) and the delivery (gauge) pressure is 200 bar or 2900 lb/in$^2$.

(16 kW, 21.4 hp.)

2 The power produced by a hydraulic motor through which oil flows at the rate of 1 l/s (61 in$^3$/s) is 5 kW or 6.7 hp. Estimate the pressure drop across the motor.

(50 bar, 725 lb/in$^2$.)

3 The oil pressure upstream of an orifice is 150 bar and downstream 70 bar. Estimate the oil velocity through the orifice.

(135.6 m/s.)

4 The pressure drop across a 6 mm (0.236 in) diameter orifice is 69 bar (1000 lb/in$^2$). Estimate the flowrate through the orifice.

($2\frac{1}{4}$ l/s, 135 in$^3$/s.)

5 Oil passes through a relief valve into a tank where all kinetic energy is dissipated. Estimate the rise in oil temperature which occurs given that the pressure drop across the valve is 113 bar (1638.5 lb/in$^2$) and that the oil specific heat is 1.67 kJ/kg degC or 0.4 Btu/lb degF.

(7.75 degC, 14 degF.)

6 Oil flows at the rate of 1 l/s (61 in$^3$/s) through a pipe of internal diameter 20 mm (0.787 in). Determine the Reynolds number if the oil viscosity is 30 cSt.

(2120.)

7 The pressure drop along a 1.5 m (59 in) length of 3 mm (0.118 in) bore tube is 50 bar (725 lb/in$^2$). Estimate the flowrate through the tube and the Reynolds number if the oil has viscosity 50 cSt.

(0.152 l/s, 9.3 in$^3$/s, (*Re*) = 1293.)

8 Oil flows through a pipe of 10 mm (0.394 in) internal diameter at the rate of 0.4 l/s (22.4 in$^3$/s). Estimate the pressure drop along a 20 m (787 in) length given the oil viscosity is 10 cSt.

(8.54 bar, 124 lb/in$^2$.)

9 Oil flows along the annular passage formed between a 10 mm
(0.394 in) diameter piston of length 20 mm (0.787 in) and its bore with a
*radial* clearance between the two of 0.05 mm (0.001 97 in). Estimate the
leakage flowrate along the annulus if the pressure difference is 200
bar (2900 lb/in$^2$) and the oil velocity is 50 cSt.

(7.5 cm$^3$/s, 0.0075 l/s, 0.46 in$^3$/s.)

10 2 l (122 in$^3$) of oil in a container has its pressure increased from 5 to
205 bar. Estimate the volume of oil which must be pumped into the
container (a) assuming the container to be rigid and (b) assuming the
container itself to increase in volume by 0.01% for every bar increase
in pressure.

(0.024 l, 1.44 in$^3$; 0.064 l, 3.9 in$^3$.)

11 Oil contains 2% of air by volume and its pressure is continually changing
between 150 and 100 bar. Estimate the effective bulk modulus assuming
a rigid container.

(4.56 × 10$^8$ N/m$^2$, 66 250 lb/in$^2$.)

12 A cylinder of 75 mm (2.95 in) internal diameter and 4 mm (0.157 in)
wall thickness is made of a metal with Young's modulus 20 × 10$^{10}$ N/m$^2$
(29 × 10$^6$ lb/in$^2$) and Poisson ratio $\frac{1}{3}$. Estimate the effective bulk modulus
of the oil in the container (a) in the absence of air and (b) with 0.1%
air present by volume if the average pressure is 250 bar or 3625 lb/in$^2$.

(14.7 × 10$^8$ N/m$^2$, 212 700 lb/in$^2$; 13.9 × 10$^8$ N/m$^2$, 200 000 lb/in$^2$.)

13 A copper pipe of 22 mm internal diameter and 1.5 mm wall thickness
contains water of specific gravity 1 and bulk modulus 21 × 10$^8$ N/m$^2$.
Estimate the speed at which a 'water hammer' surge will travel along
the pipe.

(1308 m/s, 4290 ft/s.)

14 A rigid pipeline of length 4 m containing oil has one end blocked and the
other open. Estimate the frequency in hertz of a pressure wave in the
pipe.

(About 75 Hz.)

15 A continuously operating circuit has a pump driven by an electric motor
with an adjustable relief valve to bleed excess oil through a water cooler

to a tank. The remaining fluid is fed to a motor driving a resisting load. Estimate the electric motor input power and the necessary part load cooling water flowrate for a water temperature rise of 7 degC for a unit which is designed to give a maximum 27 hp at the hydraulic motor when running at 1000 rev/min but is actually operating with a hydraulic motor power of 16 hp at 800 rev/min.

(15 kW plus losses; about 0.1 l/s.)

# 2 Dynamic Analysis

The response of hydraulic systems and servomechanisms can be much quicker than the response of their electrical or other counterparts. This is one of their advantages. For example, a hydraulic motor can accelerate or change its speed more quickly than an electric motor of the same torque rating (mainly because it has a far lower inertia). The response of a high-performance hydraulic system is usually measurable in terms of milliseconds with the permissible degree of oscillation closely restricted.

Concepts of response are based on the subject of dynamics and this chapter summarises the linear dynamic analysis of 'first-order' and 'second-order' systems and explains their responses to 'step', 'ramp' and 'harmonic' inputs. The treatment is biased towards mechanical or hydraulic components but the method is equally applicable to almost all types of system and the first example given has no serious connection with fluid power.

## 2.1  First-order Systems

Dynamic analyses of the simplest type occur with the so-called 'first-order systems'. As a homely example, consider a bottle of wine brought from a cool cellar into a warm room. The temperature of the bottle and its contents rise rapidly to begin with but more slowly as time goes on. The rate $Q$ of heat transfer from the room to the bottle depends at any instant on the temperature difference between the room and the bottle at that instant, or applying Newton's law (with its simplifying assumption that $Q$ is directly proportional to the temperature difference) we may write

$$Q = h(\theta_1 - \theta_2) \tag{2.1}$$

(where $\theta_1$ is the room temperature, $\theta_2$ the bottle temperature and $h$ a heat transfer coefficient).

This rate of heat transfer will determine the rate of rise of temperature of the bottle and its contents or

$$Q = C \frac{d\theta_2}{dt} \tag{2.2}$$

(where $C$ is the heat capacitance of the bottle and its contents).

Equating equations 2.1 and 2.2

$$C \frac{d\theta_2}{dt} = h(\theta_1 - \theta_2)$$

or

$$\theta_2 + T\frac{\mathrm{d}\theta_2}{\mathrm{d}t} = \theta_1$$

a first-order linear differential equation with a constant coefficient $T\,(=C/h)$ which has the dimensions of time (in this case Btu/degF divided by Btu/s degF, for example). $T$ is the *time constant*.

Objections to this analysis are that the temperature of the bottle is not necessarily the same as that of the wine at any instant, that the wine near the top of the bottle may be warmer than that near the bottom and that neither the heat transfer coefficient $h$ nor the heat capacity $C$ is necessarily constant. Similar objections apply to most linear dynamic analyses and the simplifying assumptions must be defined. Nonetheless, this type of analysis is used extensively because analytical results can be readily obtained and such results are found to correlate satisfactorily with experiments.

### 2.1.1 *A First-order Fluid System*

Figure 2.1 illustrates a vessel with a restricted inlet pipeline which can be regarded as first-order system with respect to the pressure in the vessel.

Assume the vessel to be initially full of fluid at pressure $P_2$ and consider the effect of a change in the pressure $P_1$ (for example, $P_1$ may suddenly rise above $P_2$).

Assume for the restrictor tube that the volume flowrate $q$ through it is directly proportional to the instantaneous pressure difference between the ends or

$$q = k(P_1 - P_2) \text{ written } k(\theta_1 - \theta_2) \tag{2.1a}$$

(for methods of estimating $k$, see chapter 1, particularly equation 1.3).

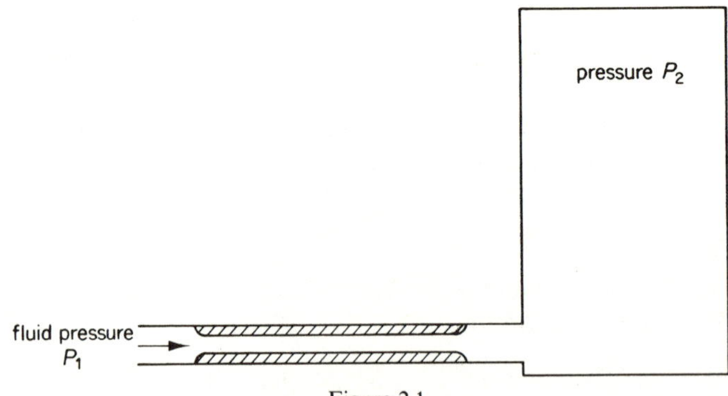


Figure 2.1

Assume that the rate of pressure rise in the vessel is directly proportional to the rate of inflow or

$$q = C \frac{\mathrm{d}P_2}{\mathrm{d}t} \quad \text{written} \quad C \frac{\mathrm{d}\theta_2}{\mathrm{d}t} \tag{2.2a}$$

(where $C$ is a 'capacitance' equal to the volume of fluid in the vessel divided by the effective bulk modulus—see equation 1.6).

Equating 2.1a with 2.2a

$$C \frac{\mathrm{d}P_2}{\mathrm{d}t} = k(P_1 - P_2)$$

or

$$P_2 + T \frac{\mathrm{d}P_2}{\mathrm{d}t} = P_1 \quad \text{written} \quad \theta_2 + T \frac{\mathrm{d}\theta_2}{\mathrm{d}t} = \theta_1$$

the same equation as before where $T(= C/k)$ has the dimensions of time. In this case $\mathrm{m}^3/(\mathrm{N}/\mathrm{m}^2)$ divided by $(\mathrm{m}^3/\mathrm{s})/(\mathrm{N}/\mathrm{m}^2)$. In British units this is $\mathrm{in}^3/(\mathrm{lb}/\mathrm{in}^2)$ divided by $(\mathrm{in}^3/\mathrm{s})/(\mathrm{lb}/\mathrm{in}^2)$, for example.

### 2.1.2  *A First-order Electrical System*

Figure 2.2 shows a simple *RC* circuit, another first-order system as regards the voltage across the condenser.

If $e_1$ (or $\theta_1$) is the voltage applied to the circuit and $e_2$ (or $\theta_2$) is the potential difference across the condenser at any instant, the voltage across the resistance will determine the current at that instant or

$$I = \frac{1}{R}(e_1 - e_2) \text{ written } \frac{1}{R}(\theta_1 - \theta_2) \tag{2.1b}$$

The rate of build-up of charge on the condenser will equal the current and from the definition of capacitance

$$I = C \frac{\mathrm{d}e_2}{\mathrm{d}t} \quad \text{written} \quad C \frac{\mathrm{d}\theta_2}{\mathrm{d}t} \tag{2.2b}$$

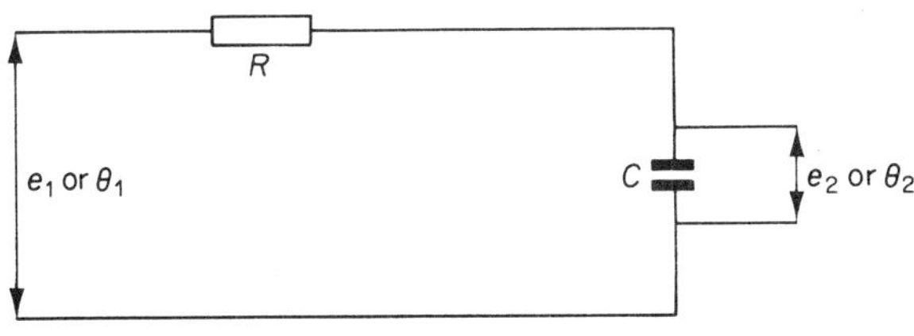


Figure 2.2

Equating 2.1b with 2.2b

$$C\frac{\mathrm{d}\theta_2}{\mathrm{d}t} = \frac{1}{R}(\theta_1 - \theta_2)$$

or

$$\theta_2 + T\frac{\mathrm{d}\theta_2}{\mathrm{d}t} = \theta_1$$

a first-order equation as before with $T(=RC)$ having the dimensions of time (in this case (volts/amperes) × (coulombs/volts) and coulombs equal ampere seconds).

### 2.1.3 *A First-order Hydraulic Servomechanism*

Figure 2.3 illustrates a simple hydraulic device which can, *under special circumstances*, be regarded as a first-order system (see chapter 8 for more detailed analysis).

 A displacement of the input horizontally causes the valve to open admitting fluid to one side of the piston and exhausting fluid from the other, thereby moving the piston. This piston movement causes the valve to return gradually to its closed position. Movements of the piston follow movements of the input. If the input is suddenly displaced and then held stationary, the effect is for the piston to start moving and then to continue moving progressively more slowly until the valve is shut with the piston in a new position. The force at the input need only be sufficient to move the valve whereas the force at the piston can be thousands times greater.

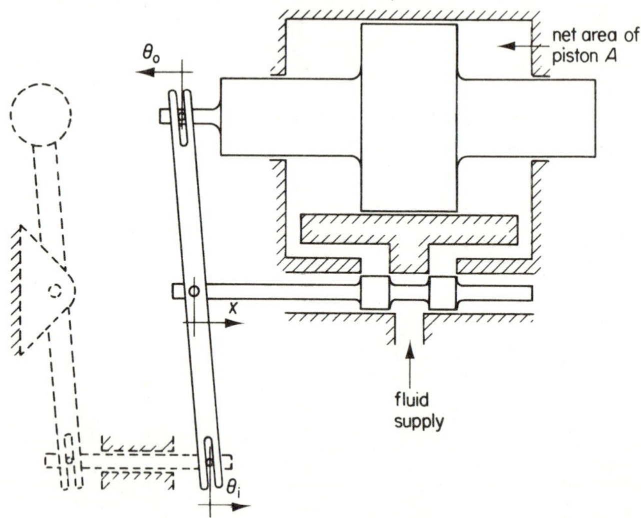


Figure 2.3 Hydraulic relay: valve displacement $x$; input displacement $\theta_i$; output displacement $\theta_o$

If the valve pivot is half-way along the feedback lever, valve displacements $x$ are given by

$$x = \tfrac{1}{2}(\theta_i - \theta_o)$$

(a different geometrical ratio applying to different lever ratios). The first approximation involved here is the assumption that distances between pivot points and fulcrum do not change owing to angularity of the linkage.

Assume the flowrate $q$ through either side of the valve is directly proportional to valve displacement or

$$q = kx \tag{2.1c}$$

which is a serious oversimplification (see chapter 7).

Assuming (a) that there is zero load on the piston rod (that is no friction, no inertia, no external forces), (b) that there is no leakage and (c) that the fluid is incompressible, this gives piston velocity directly proportional to flowrate or

$$q = A\frac{\mathrm{d}\theta_o}{\mathrm{d}t} \tag{2.2c}$$

an unrealistic set of conditions but given here because the first-order analyses can be used for approximate calculations and so as to illustrate how a first-order approximation can be derived.

Equating 2.1c and 2.2c

$$A\frac{\mathrm{d}\theta_o}{\mathrm{d}t} = \frac{k}{2}(\theta_i - \theta_o)$$

or

$$\theta_o + T\frac{\mathrm{d}\theta_o}{\mathrm{d}t} = \theta_i$$

where $T(= 2A/k)$ has the dimensions of time ($\mathrm{m}^2$ divided by $(\mathrm{m}^3/\mathrm{s})/\mathrm{m}$ for example ).

### 2.1.4  *The First-order Equation*

All first-order systems have a governing equation of the form

$$\theta_2 + T\frac{\mathrm{d}\theta_2}{\mathrm{d}t} = \theta_1 \tag{2.3}$$

which may be written in operator notation (with $\mathrm{D} \equiv \mathrm{d}/\mathrm{d}t$)

$$\theta_2 + T\mathrm{D}\theta_2 = \theta_1 \tag{2.4}$$

and the convention is to rewrite this equation as

$$\frac{\theta_2}{\theta_1} = \frac{1}{1 + T\mathrm{D}} \tag{2.5}$$

Equations 2.3, 2.4 and 2.5 are merely alternative forms of the same relation. (In some cases a first-order equation takes the form

$$\frac{\theta_2}{\theta_1} = \frac{K}{1 + T\mathrm{D}}$$

where $K$ is some constant and called the steady state gain.) First-order systems are sometimes called 'single-capacity systems'.


## 2.2   The Step Input

In all the cases considered above, $\theta_1$ (or $\theta_i$) is the demand or 'input' value whereas $\theta_2$ (or $\theta_o$) is the actual or 'output' value (number suffixes are used for component parts or subsystems whereas letter suffixes are used for complete servo systems). In the hydraulic system of figure 2.3 an input displacement $\theta_i$ causes a piston displacement $\theta_o$.

If the input of the hydraulic system is suddenly displaced horizontally distance $N$, and thereafter held still, the system is said to have been subjected to a step input. Similarly, if a constant voltage is suddenly applied to the electrical circuit of figure 2.2 or if there is a sudden rise in the pressure $P_1$ for the fluid circuit of figure 2.1, step inputs are said to have been applied.

A sudden change from some datum which is taken as zero to some other

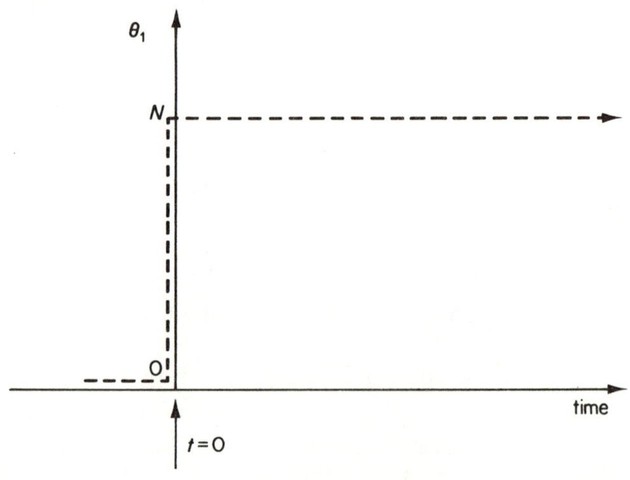


Figure 2.4 The step input

value (say $N$) is termed a 'step' and may be illustrated as in figure 2.4 or written

$$\theta_1 = 0 \text{ for } t = 0^-$$

$$\theta_1 = N \text{ for } t = 0^+$$

### 2.2.1  *Response of First-order Systems to Step Input*

Applying a step input of magnitude $N$ to the hydraulic system of figure 2.3 causes the piston to start moving if the system is quiescent before the step is applied. After sufficient time has elapsed, the piston will have moved distance $N$ from its original position and a new quiescent state will have been reached. The piston and valve then remain stationary until the input is next disturbed.

For the electrical circuit of figure 2.2, a sudden application of $N$ volts causes the voltage across the condenser to begin rising until it finally reaches $N$ volts. For the fluid system of figure 2.1, the initial pressure in the vessel and supply line might be $P_0$ (taken as zero). A step increase at the inlet of $N$ (units of pressure) causes the container pressure to start rising and it finally reaches the new pressure $P_0 + N$.

### 2.2.2  *Response as a Function of Time*

The output (or value of $\theta_2$ or $\theta_o$) finally attained is obvious but dynamic analysis concerns instantaneous values and, for first-order systems, the instantaneous value of $\theta_2$ (or $\theta_o$) at any time $t$ following a step input of magnitude $N$ is given by

$$\theta_2 = N\{1 - \exp(-t/T)\} \tag{2.6}$$

(step response, first-order system). The plot of equation 2.6 is given in figure 2.5 and shows $\theta_2$ exponentially approaching the value $N$. Note that, at time $T$ after applying the step, the value of $\theta_2$ is 0.631 (namely $1 - e^{-1}$)

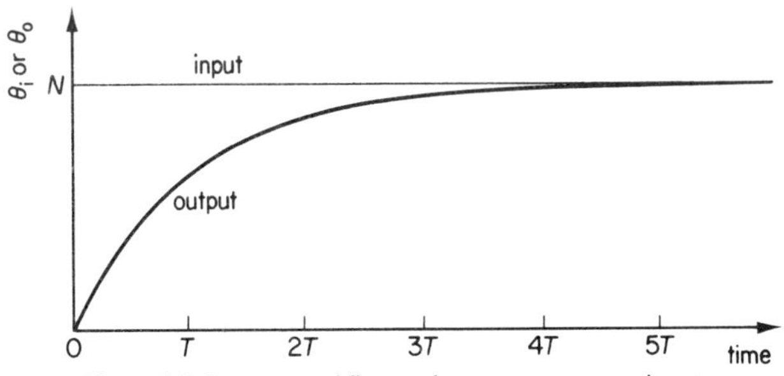


Figure 2.5  Response of first-order systems to step input

times the magnitude of the step. Also, by time $5T$, the output will be within $1\%$ of its final value $(1 - e^{-5} \approx 99\%)$.

Equation 2.6 is the solution of equation 2.5 (or 2.4 or 2.3) for $\theta_1$ a step of magnitude $N$. The solution can be verified experimentally or obtained mathematically.

Note that $1 + T\lambda = 0$ is termed the subsidiary equation of a first-order system and that the root of this equation ($\lambda = -1/T$) determines the response.

First-order systems are sometimes called 'simple exponential delays'.

## 2.3 Ramp Input and Response for First-order Systems

A ramp input means, for a position control system, that the input suddenly starts moving at steady velocity (say $\Omega$) and, for other types of device, that the input suddenly starts increasing at a constant rate $\Omega$. The ramp input may be written $\theta_1$ (or $\theta_i$) = 0 for $t = 0^-$; $\theta_1$ (or $\theta_i$) = $\Omega t$ for $t = 0^+$, where $\Omega$ is a constant.

A sufficient length of time after applying a ramp input to a first-order system (for example after at least $5T$), the output will be moving at the same velocity (or more generally increasing at the same rate) as the input but will be lagging behind the input (for example in position) by an amount $\Omega T$ as illustrated in figure 2.6.

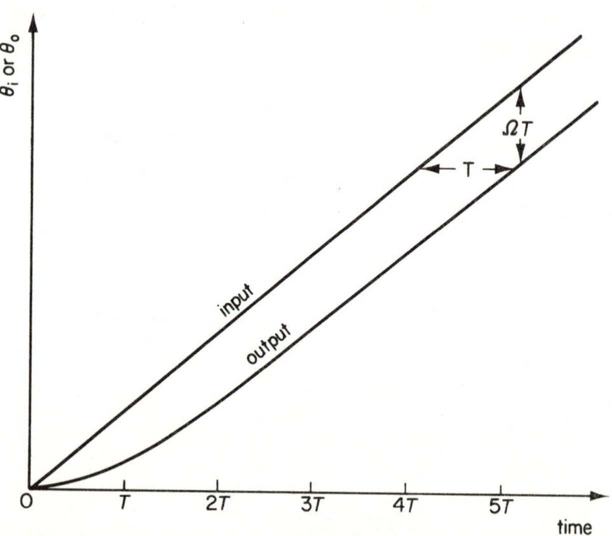


Figure 2.6  Response of first-order systems to ramp input

The solution of equation 2.5 for $\theta_1 = \Omega t$ on which figure 2.6 is based, may be written

$$\theta_2 = \Omega t - \Omega T \{1 - \exp(-t/T)\} \tag{2.7}$$

(ramp response, first-order system).

The 'following error' or 'velocity lag' for a first-order system equals $\Omega T$.

## 2.4  Harmonic Input

Referring to the electrical circuit of figure 2.2, if an a.c. voltage is being applied, then the circuit is said to have a simple harmonic input (at some frequency $\omega$). The output (in this case the voltage across the condenser) will also be an a.c. voltage of the same frequency as the input but with a different amplitude and out of phase with the input.

Similarly with the hydraulic device of figure 2.3; if the input is being moved back and forth sinusoidally, the system is being subjected to a harmonic input. The output (in this case the piston motion) will also be simple harmonic at the same frequency but with a different amplitude and out of phase with the input. In the case of this hydraulic system an experiment might show a distorted output waveform but for analysis it is assumed to be a pure sine wave.

The harmonic or sine wave is assumed to be going on continuously—unlike other types of input, we are not interested in the starting-up period of a harmonic. The simple harmonic input is usually considered to have unit amplitude and may be represented as

$$\theta_1 \,(\text{or } \theta_i) = 1 \sin(\omega t) \,(\text{or } \theta_1 \,(\text{or } \theta_i) = 1 \cos(\omega t))$$

(where $\omega$ is the particular frequency being considered—$\omega$ has units rad/s). But it is mathematically more convenient to write the unit amplitude harmonic as

$$\theta_1 \,(\text{or } \theta_i) = \exp(i\omega t)$$

(where the real part represents the cosine and the imaginary part the sine).

### 2.4.1  *Harmonic Response of First-order Systems*

The output of any linear system (of which first-order systems are the simplest) which is subjected to a simple harmonic input is also a simple harmonic at the same frequency with its amplitude depending on the frequency. For the particular case of first-order systems, the amplitude of the output is always less than that of the input. Figure 2.7 illustrates qualitatively the output with respect to time for three different frequencies.

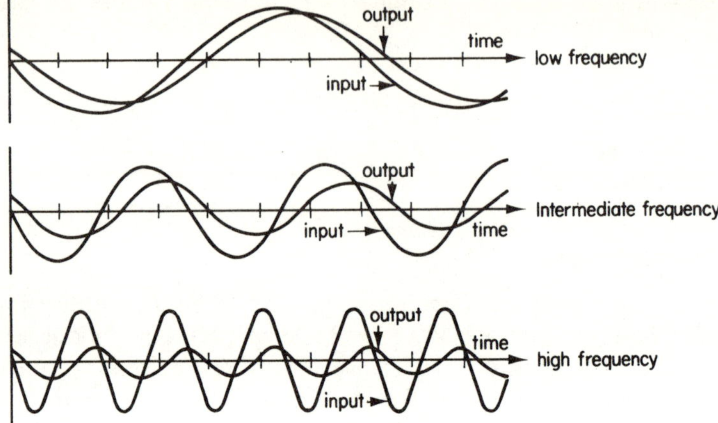


Figure 2.7  Diagram illustrating harmonic response

Mathematically, the 'steady state' solution (or particular integral) of equation 2.5 (or 2.4 or 2.3) when $\theta_1 = \mathrm{Im}\{\exp(i\omega t)\}$ or $1\sin(\omega t)$ is

$$\theta_2 = \mathcal{M}\sin(\omega t - \phi) \tag{2.8}$$

where

$$\mathcal{M} = \frac{1}{(1 + \omega^2 T^2)^{1/2}} \text{ and } \tan\phi = \omega T$$

(harmonic response, first-order system) where $\mathcal{M}$ is the amplitude ratio (of output to input) or the 'dynamic magnification' (which for the special case of first-order systems is always less than unity). Also $\phi$ is the 'phase angle' or 'phase lag' (which for first-order systems is always less than 90°).

### 2.4.2  *Graphical Representations*

The term $\exp(i\omega t)$ can be considered as a vector of unit length which is rotating anticlockwise with angular velocity $\omega$ radians per second, its projection on the imaginary axis plotted against time being a sine wave. The output is known to be $\mathcal{M}\sin(\omega t - \phi)$ which can also be considered as a vector rotating in the same sense and with the same angular velocity but having a length $\mathcal{M}$ and lagging the input by $\phi$. For a particular frequency $\omega_1$, with amplitude ratio $\mathcal{M}_1$ and phase angle $\phi_1$, the two vectors and their projections appear as in figure 2.8.

For convenience the term $\exp(i\omega t)$ (or $1\sin(\omega t)$) is plotted along the real positive axis as a stationary vector of unit length. Then the output for frequency $\omega_1$ becomes a stationary vector of length $\mathcal{M}_1$ at angle $\phi_1$ clockwise from the unit vector as shown in figure 2.9.

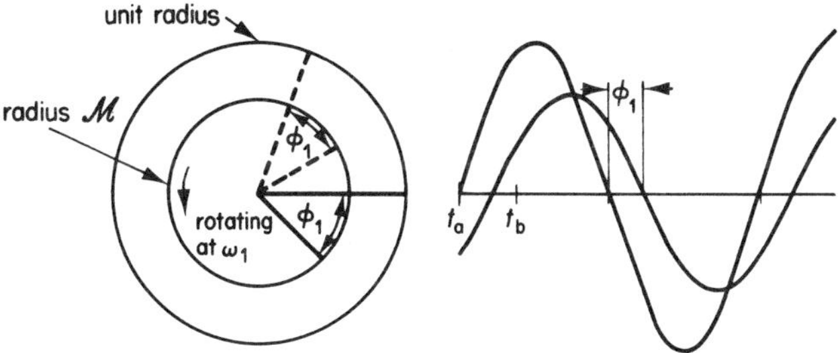



Figure 2.8 Harmonic vectors:——vectors at time $t_a$;---vectors at time $t_b$

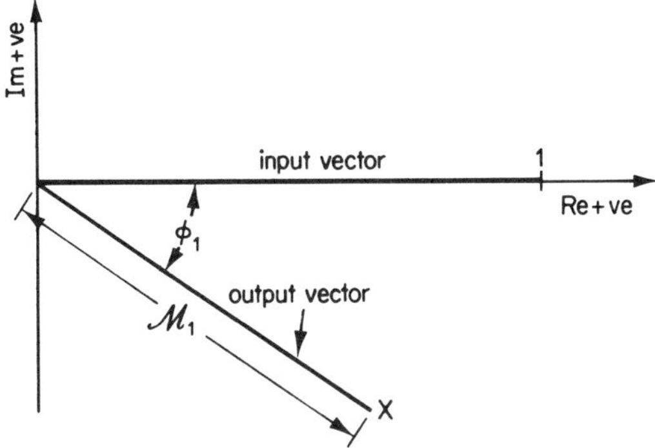



Figure 2.9 Output vector for one frequency

The end point $X$ of the output vector shown in figure 2.9 is defined by $1/(1 + Ti\omega_1)$ (the projection on the real axis being $1/(1 + \omega_1{}^2 T^2)$ and on the imaginary axis $-\omega_1 T/(1 + \omega_1{}^2 T^2)$ and merely represents an alternative way of stating equation 2.8. Thus we can take equation 2.5 and substitute $i\omega$ for D in order to obtain the harmonic relation

$$\frac{\theta_2}{\theta_1} = \frac{1}{1 + i\omega T} \quad \text{for } \theta_1 = \exp(i\omega t)$$

and a more general relation can also be written as

$$\frac{\theta_2}{\theta_1} = \frac{1}{1 + Ts} \quad \text{or} \quad \frac{1}{1 + Tp}$$

where $s$ or $p$ represent the complex operator or Laplace operator when the expression is then called the 'transfer function'.

The term $\exp(i\omega t)$ is represented as a stationary vector of unit length lying along the real positive axis whatever the frequency may be. The output vector, however, changes as regards length $\mathcal{M}$ and phase angle $\phi$ as the frequency changes. Three representative output vectors for three different

                    *Hydraulic Systems Analysis*

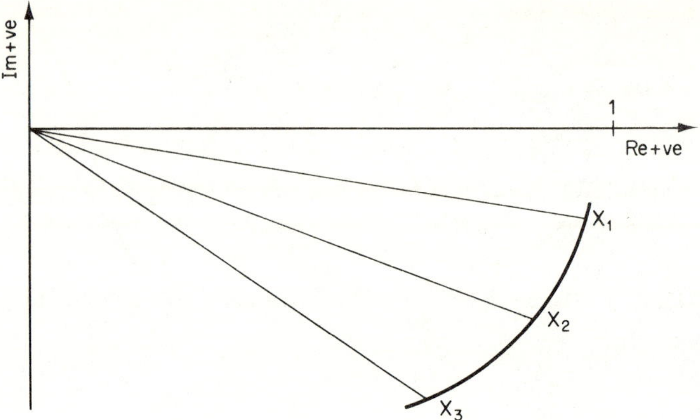

Figure 2.10 Output vectors for three different frequencies $(\omega_1 < \omega_2 < \omega_3)$

frequencies $(\omega_1 < \omega_2 < \omega_3)$ are shown in figure 2.10 with their end points $X_1, X_2$ and $X_3$ joined together.

Note that a frequency of particular interest with first-order systems is $\omega = 1/T$ when $\phi = 45°$ and $\mathcal{M} = 1/(2)^{1/2}$.

### 2.4.3  *Harmonic Response Locus*

The harmonic response locus is the line joining the end points of all output vectors for all frequencies between $\omega = 0$ and $\omega = \infty$. In the case of all first-order systems, the locus is a semicircle as shown on figure 2.11.

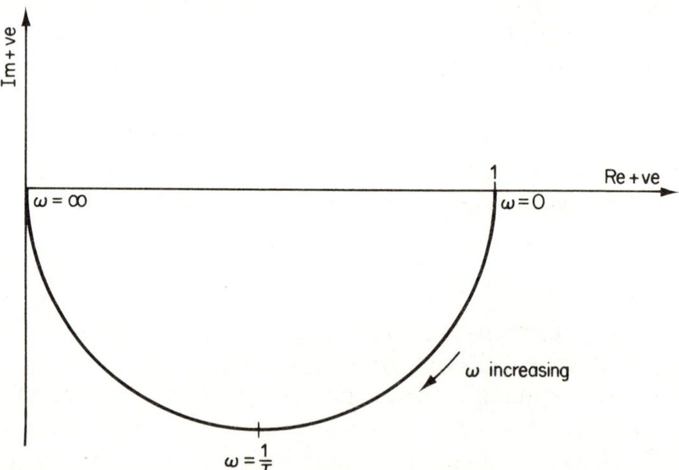

Figure 2.11  Harmonic response locus for first-order systems

### 2.4.4 *Logarithmic Plots*

A popular method of representing harmonic response, because of its usefulness in dealing with complicated systems, is to plot amplitude ratio versus frequency on logarithmic scales. The scales are decibels and octaves respectively. Amplitude ratio in decibels is *defined* as $20 \log_{10} \mathscr{M}$ : thus, if $\mathscr{M} = 2$, the value is $+ 6$ dB ($20 \log_{10} 2 \approx 6$); and, for $\mathscr{M} = \frac{1}{2}$, the value is $- 6$ dB. The octave scale for frequency is a logarithmic scale to base 2. Doubling a selected frequency means an increase of 1 octave and halving it means a decrease of 1 octave.

The decibel–octave curve for a first-order system can be plotted from numerical evaluations of the term $1/(1 + \omega^2 T^2)^{1/2}$ (see equation 2.8) with the result shown in figure 2.12.

A close approximation to this curve can be obtained by using the asymptotic straight lines which are along the 0 dB line (for $\omega$ small, the value of $\mathscr{M}$ tends to 1) and along a line of slope $- 6$ dB/octave passing through $\omega T = 1$ at 0 dB which is defined as $- 6 \log_2 \omega T$ (because for $\omega$

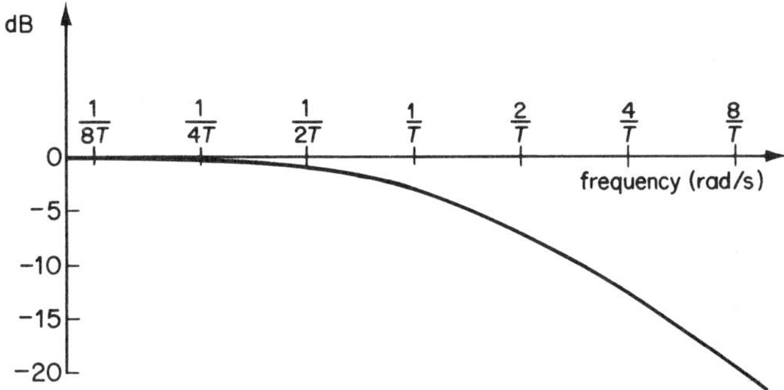

Figure 2.12 Amplitude versus frequency (first-order systems)

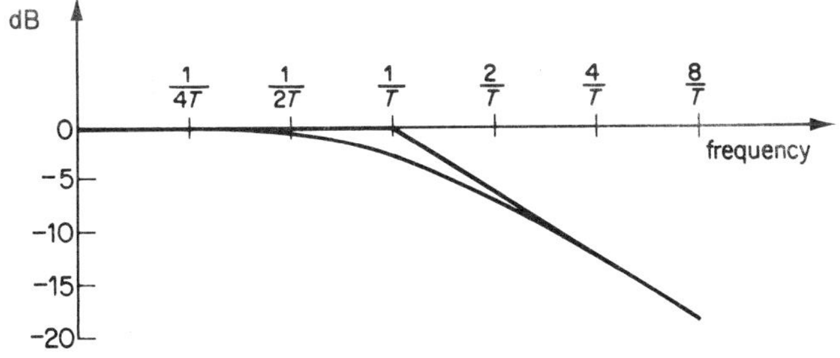

Figure 2.13 Straight-line approximation (first-order systems)

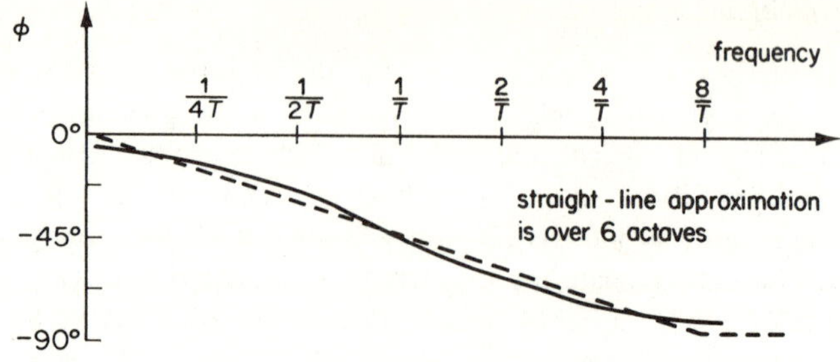



Figure 2.14  Phase angles (first-order systems)

large, the value of $\mathcal{M}$ tends to $1/\omega T$ so $+20 \log_{10} \mathcal{M} \approx -20 \log_{10} \omega T = -20 \log_2 \omega T \log_{10} 2 \approx -6 \log_2 \omega T$). The straight-line approximation is shown in figure 2.13 and deviates from the true curve by 3 dB at $\omega = 1/T$ and by 1 dB at $\omega = 2/T$ and $1/2T$. Also shown in figure 2.14 is the phase angle plot (not logarithmic) against log frequency.

## 2.5  Second-order Systems

The classic example of second-order systems consists of a mass, a spring and a viscous damper. A mass attached to the lower end of a spring and with a damper attached to it is shown in figure 2.15.

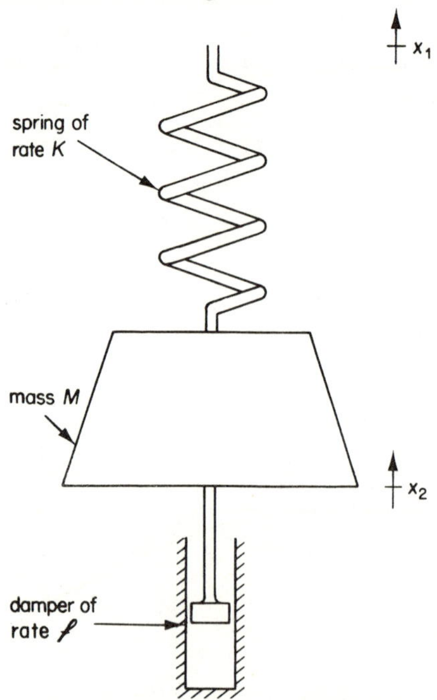



Figure 2.15  System consisting of a mass, a spring and a viscous damper

If the free end of the spring is suddenly moved vertically and then held stationary, the mass will begin to oscillate but the oscillations will decay. A general analysis concerns the relation between input displacements $x_1$ and the resulting output displacements $x_2$. It is based on the assumption that the spring and damper have no mass (that is $M$ is the only mass to be considered). It is concerned with dynamic forces because the (static) spring extension necessary to support the weight need not be considered (the force acting downwards on the mass due to gravity is exactly balanced by the force acting upwards on the mass due to the static spring extension— the quiescent state of the system defines the zero points of $x_1$ and $x_2$).

The force required to accelerate the mass is

$$M \frac{\mathrm{d}^2 x_2}{\mathrm{d}t^2}$$

The force applied to the mass by the spring is

$$K(x_1 - x_2)$$

The force applied to the mass by the damper is

$$-f \frac{\mathrm{d}x_2}{\mathrm{d}t}$$

Forces on the mass and displacements are taken as positive upwards.

The accelerating force equals the net force applied to the mass or

$$M \frac{\mathrm{d}^2 x_2}{\mathrm{d}t^2} = K(x_1 - x_2) - f \frac{\mathrm{d}x_2}{\mathrm{d}t}$$

or

$$M \frac{\mathrm{d}^2 x_2}{\mathrm{d}t^2} + f \frac{\mathrm{d}x_2}{\mathrm{d}t} + K x_2 = K x_1$$

or

$$\frac{M}{K} \mathrm{D}^2 x_2 + \frac{f}{K} \mathrm{D}x_2 + x_2 = x_1$$

The analysis gives a second-order linear differential equation with constant coefficients (the coefficients being $M/K$ and $f/K$).

The expression $(K/M)^{1/2}$ or $(1/2\pi)(K/M)^{1/2}$ is known to define the natural frequency (in rad/s and Hz respectively) of an undamped mass spring system and the first coefficient $M/K$ of the second-order equation may be written as

$$\frac{1}{\omega_n^{\,2}}$$

($\omega_n$ is the natural frequency of a system having a mass of the same magnitude $M$ supported by a massless spring of the same rate $K$ with no damper present).

The second coefficient $f/K$ is conveniently dealt with by introducing the 'damping ratio' or 'damping factor' $\zeta$ by writing

$$\frac{f}{K} = \frac{2\zeta}{\omega_n}$$

The second-order equation in more general form (substituting $\theta_1$ for $x_1$ and $\theta_2$ for $x_2$) thus becomes

$$\frac{1}{\omega_n^2} D^2\theta_2 + \frac{2\zeta}{\omega_n} D\theta_2 + \theta_2 = \theta_1$$

### 2.5.1    *A Second-order Electrical Circuit*

Figure 2.16 shows an $\mathscr{L}$RC circuit which is also a second-order system.

If $e_{\mathscr{L}}$, $e_R$ and $e_2$ are the potential differences across the inductance, the resistance and the capacitance respectively at any instant and the potential being applied is $e_1$, then

$$e_{\mathscr{L}} + e_R + e_2 = e_1$$

and, if $I$ is the current at any instant, we have

$$e_{\mathscr{L}} = \mathscr{L}\frac{dI}{dt} \quad \text{and} \quad I = C\frac{de_2}{dt}$$

giving

$$\frac{dI}{dt} = C\frac{d^2e_2}{dt^2} \quad \text{or} \quad e_{\mathscr{L}} = \mathscr{L}C\frac{d^2e_2}{dt^2}$$

also

$$e_R = RI = RC\frac{de_2}{dt}$$

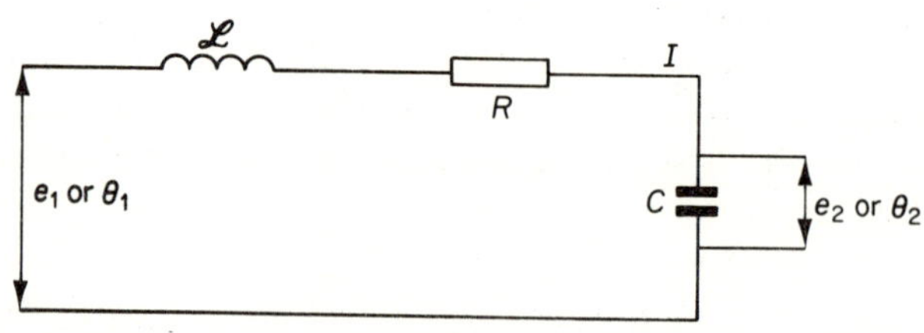


Figure 2.16

Hence

$$\mathscr{L}C\,\frac{\mathrm{d}^2e_2}{\mathrm{d}t^2} + RC\,\frac{\mathrm{d}e_2}{\mathrm{d}t} + e_2 = e_1$$

or, substituting $\theta_2$ for $e_2$, $\theta_1$ for $e_1$ and D for $\mathrm{d}/\mathrm{d}t$,

$$\mathscr{L}C\,\mathrm{D}^2\theta_2 + RC\,\mathrm{D}\theta_2 + \theta_2 = \theta_1$$

and an $\mathscr{L}C$ circuit with no resistance has natural frequency of $1/(\mathscr{L}C)^{1/2}$, so equation may be written

$$\frac{1}{\omega_\mathrm{n}^{\,2}}\,\mathrm{D}^2\theta_2 + \frac{2\zeta}{\omega_\mathrm{n}}\,\mathrm{D}\theta_2 + \theta_2 = \theta_1$$

the same equation as before (note that the $Q$ factor in electrical terminology is given by $Q = 1/2\zeta$).

### 2.5.2   The Second-order Equation

All second-order systems have a governing equation of the form

$$\frac{1}{\omega_\mathrm{n}^{\,2}}\,\mathrm{D}^2\theta_2 + \frac{2\zeta}{\omega_\mathrm{n}}\,\mathrm{D}\theta_2 + \theta_2 = \theta_1 \tag{2.9}$$

or

$$\mathrm{D}^2\theta_2 + 2\zeta\omega_\mathrm{n}\,\mathrm{D}\theta_2 + \omega_\mathrm{n}^{\,2}\theta_2 = \omega_\mathrm{n}^{\,2}\theta_1$$

which is conventionally written

$$\frac{\theta_2}{\theta_1} = \frac{1}{(1/\omega_\mathrm{n}^{\,2})\mathrm{D}^2 + (2\zeta/\omega_\mathrm{n})\mathrm{D} + 1} \tag{2.10}$$

(In some cases a second-order relation takes the form

$$\frac{\theta_2}{\theta_1} = \frac{K}{(1/\omega_\mathrm{n}^{\,2})\mathrm{D}^2 + (2\zeta/\omega_\mathrm{n})\mathrm{D} + 1}$$

where $K$ is the steady state gain.)

Applying a step input of magnitude $N$ to either the mechanical system of figure 2.15 or the electrical system of figure 2.16 causes the output to begin changing, assuming the system is quiescent before the step is applied. After sufficient time has elapsed the mass will have moved a distance $N$ or the condenser voltage increased by an amount $N$ and a new quiescent condition will have been reached.

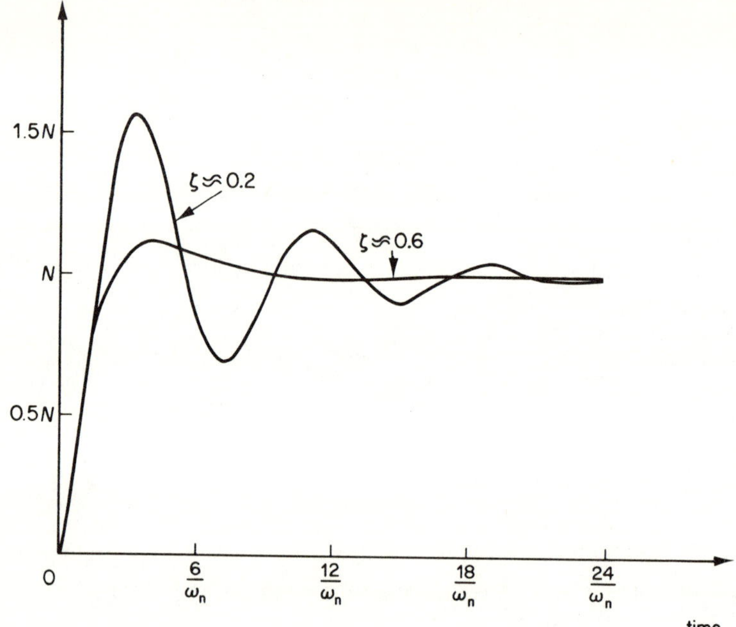


Figure 2.17 Response of second-order systems to step input

The instantaneous value of the output at any time $t$ after applying the step will be given (for any value of $\zeta$ less than 1) by the following equation:

$$\theta_2 = N - N \frac{1}{(1 - \zeta^2)^{1/2}} \exp(-\zeta\omega_n t) \sin(bt + \phi) \tag{2.11}$$

where

$$b = \omega_n(1 - \zeta^2)^{1/2} \text{ and } \tan\phi = \frac{(1 - \zeta^2)^{1/2}}{\zeta}$$

(step response, second-order systems).

Plots of equation 2.11 versus time are given in figure 2.17 and show that the output $\theta_2$ first overshoots the desired value $N$ and finally settles to this value. The amount of overshoot and (to a lesser extent) the frequency of the induced oscillations vary with $\zeta$.

Note that

$$\lambda^2 + 2\zeta\omega_n\lambda + \omega_n{}^2 = 0$$

is the subsidiary equation of a second-order system whose roots determine the response.

(The roots are

$$\lambda_1, \lambda_2 = \frac{-2\zeta\omega_n \pm (4\zeta^2\omega_n{}^2 - 4\omega_n{}^2)^{1/2}}{2}$$

or

$$-\omega_n\zeta \pm i\omega_n(1 - \zeta^2)^{1/2}$$

and a term

$$\exp(-\omega_n\zeta t)(C_1 \exp[\{i\omega_n(1 - \zeta^2)^{1/2}\}t] + C_2 \exp[\{i\omega_n(1 - \zeta^2)^{1/2}\}t])$$

can be written

$$\frac{1}{(1 - \zeta^2)^{1/2}} \exp(-\zeta\omega_n t) \sin[\{\omega_n(1 - \zeta^2)^{1/2}\}t + \phi]$$

as above.) Second-order systems are sometimes called 'two-capacity systems' (that is they have two forms of energy storage).

For heavily damped systems with $\zeta > 1$ (more than the so-called 'critical' damping), no oscillations occur following a step input (the response is aperiodic) and there is no overshoot.

## 2.6 Response of Second-order Systems to Ramp Input

As with first-order systems, the output for a ramp input will finally attain the same velocity (or rate of change) as the input but there will be a lag. This 'velocity lag' or 'following error' will equal

$$\frac{2\zeta}{\omega_n}\Omega$$

where $\Omega$ is the input velocity (or rate of change). Also, for $\zeta < 1$, the approach to this final velocity will be oscillatory.

The relation (that is the solution of equation 2.10 or 2.9 for $\theta_1$ a ramp input) is

$$\theta_2 = \Omega t - \Omega\frac{2\zeta}{\omega_n} + \exp(-\zeta\omega_n t)\frac{\Omega}{(1 - \zeta^2)^{1/2}} \sin[\{\omega_n(1 - \zeta^2)^{1/2}\}t + \phi] \tag{2.12}$$

where

$$\tan\frac{\phi}{2} = \frac{(1 - \zeta^2)^{1/2}}{\zeta}$$

(response to ramp input, second-order systems).

## 2.7 Harmonic Response of Second-order Systems

The output of a second-order system (as with all linear systems) subjected to a simple harmonic input of frequency $\omega$ is also a simple harmonic at the same frequency but of different amplitude and out of phase from the input.

Mathematically, the 'steady state' solution of equation 2.10 (or 2.9) for $\theta_1 = \mathrm{Im}\,\{\exp(i\omega t)\}$ is given by

$$\theta_2 = \mathscr{M}\sin(\omega t - \phi) \qquad\qquad (2.13)$$

where the amplitude ratio

$$\mathscr{M} = \frac{1}{\{(1 - \imath^2)^2 + 4\zeta^2\imath^2\}^{1/2}}$$

($\imath$ being the ratio of the applied or forcing frequency $\omega$ to the natural frequency $\omega_n$: $\imath = \omega/\omega_n$) and the phase angle is obtained from $\tan\phi = 2\zeta\,\imath/(1 - \imath^2)$.

Equation 2.13 is identical with

$$\frac{\theta_1}{\theta_1} = \frac{1}{(1/\omega_n^2)(i\omega)^2 + (2\zeta/\omega_n)(i\omega) + 1} \text{ for } \theta_1 = \exp(i\omega t)$$

obtained by replacing D in equation 2.10 by $i\omega$. The transfer function for second-order systems is

$$\frac{1}{(1/\omega_n^2)s^2 + (2\zeta/\omega_n)s + 1}$$

The amplitude ratio $\mathscr{M}$ may be considerably greater than unity particularly if little damping is present (that is if $\zeta$ is small). For practical purposes, if $\zeta$ is less than about $\frac{1}{4}$, the maximum value of $\mathscr{M}$ may be taken to occur when the forcing frequency or test frequency equals the natural frequency (that is $\omega = \omega_n$ or $\imath = 1$) when $\mathscr{M} = 1/2\zeta$ and $\phi = 90°$ lagging.


### 2.7.1   *Harmonic Response Locus*

The locus for any system is the curve joining the end points of all output vectors for frequencies between 0 and $\infty$, taking the input as the unit vector. The shape of the curve for second-order systems varies with the damping ratio $\zeta$ as illustrated in figure 2.18 (on which the $\omega = \omega_n$, $90°$ points are marked X).


### 2.7.2   *Logarithmic Plots*

The decibel–octave curve for a second-order system can be approximated— but only very roughly—by straight-line asymptotes, but the amplitude at $\omega = \omega_n$ should be particularly noted (this is $-20\log_{10}2\zeta$ and equals about $+14$ dB for $\zeta = 0.1$ and $+6$ dB for $\zeta = 0.2$). The approximate plot

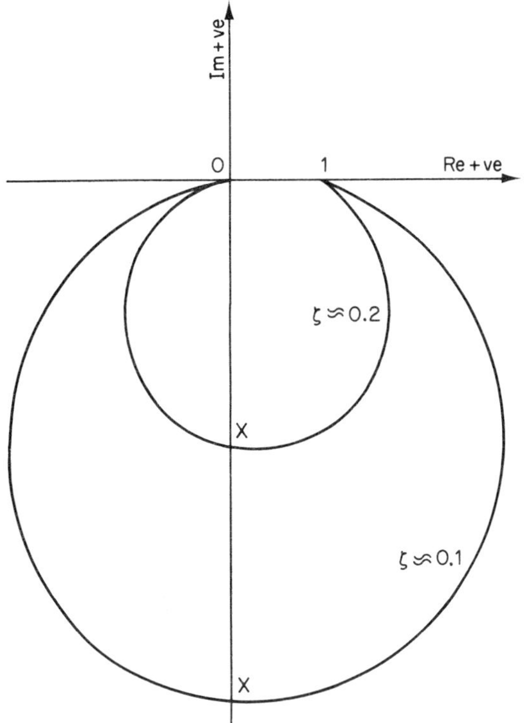



Figure 2.18 Harmonic response loci for second-order systems

comprises two straight lines intersecting at 0 dB, and $\omega = \omega_n$ with the first line along the 0 dB axis (for $\omega$ small that is $\imath \to 0$, then $\mathcal{M} \to 1$ and $20 \log_{10} \mathcal{M} \to 0$). The second line slopes at $-12$ dB/octave (double the first-order slope), because for high frequencies $\omega$ or $\imath \to \infty$, $\mathcal{M} \to 1/\imath^2$ and $20 \log_{10} \mathcal{M} \to -40 \log_2 \imath \log_{10} 2 \approx -12 \log_2 \imath$.

The resulting plot is shown as the broken line of figure 2.19 together with the actual curves (full lines) for selected values of $\zeta$. Phase angles also vary with $\zeta$ as shown.

## Problems

1 A simple ball valve controls the water level in a tank in such a way that the rate of flow into the tank is directly proportional to the vertically downwards displacement of the ball from a given level. The flow from the tank may vary and, when it is at a steady value of $1/200$ m$^3$/s, the ball is steady 1 cm below the given level. The cross-sectional area of the tank is 5 m$^2$. Making simplifying assumptions, show that, if the flow

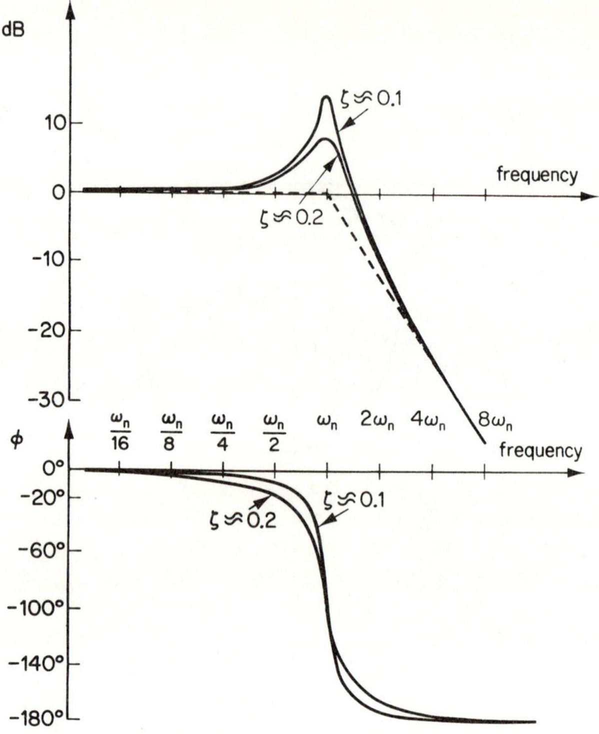



Figure 2.19  Harmonic response of second-order systems on logarithmic scales

from the tank suddenly increases from 0 to $1/200 \text{ m}^3/\text{s}$, the ball approaches the 1 cm position according to the law $h = 1 - \exp(-t/10)$ cm and that the level settles, to within 1% of 1 cm, 46 s after the flow begins. (Note. Water level may be considered as the output and potential level corresponding to the flow from the tank as the input.)

2 A thermometer for use in a control system has a phial and a scale which at all instants records the actual phial temperature. The phial initially at 0 °C is suddenly plunged into a liquid at 100 °C. Show that $1\frac{1}{2}$ s later the scale reading will be 63.2 °C if the following data apply: heat transfer coefficient $h = 20 \text{ kW/m}^2$ degK; phial specific heat $C = 1.2 \text{ kJ/kg degK}$; mass of phial $m = 0.25$ kg; surface area of phial $A = 0.01 \text{ m}^2$. Assume $h$, $C$ and liquid temperature $\theta_1$ all constants. (Hint. Obtain $T = 1\frac{1}{2} \text{ s} = mC/hA$.)

3 A hydraulic motor of negligible inertia has attached to its shaft a flywheel of inertia 5 kg m². The swept volume of the motor is 628 cm³/rev. Leakage

is directly proportional to pressure drop across the motor and occurs at the rate of 25 cm$^3$/s for every bar pressure drop. Assuming that fluid compressibility and friction are negligible, show that the motor speed exponentially approaches a steady value without overshooting after flow to the motor suddenly begins, and that the motor speed is $0.632 \times 250$ rev/min, $\frac{1}{8}$ s after the start of a flow of 2620 cm$^3$/s. Also estimate the maximum pressure across the motor, that is flow/leakage rate—105 bars. (Hint. $\tau = \delta_m P$ for a hydraulic motor where $\tau$ is torque, motor displacement $\delta_m$/rad and $P$ pressure difference across motor; torque equals work done/rad.)

4 A hydraulic relay (see figure 8.4) has a feedback link of ratio $8:1$ (viz. 8 units between input and output pivots, 1 unit between input and spool valve pivots, 9 units between spool valve and output pivots). The flowrate through the valve is 400 cm$^3$/s for maximum spool displacement 5 mm and the ram area is $8\frac{1}{3}$ cm$^2$. Show that the relay behaves as a first-order (simple exponential delay) system with time constant $1/12$ s if the following assumptions hold: (a) fluid incompressible, (b) flowrate through valve directly proportional to spool displacement, (c) leakage and friction negligible and (d) inertia of all parts and load negligible. If the input is moving back and forth sinusoidally about the midposition with a frequency of 1 Hz ($2\pi$ rad/s), what is the phase lag and what is the time lag between the input passing the midposition and the output passing through its midposition?

(27.7°, 1/13 s approx.)

5 A mass $M$ is supported by the lower end of a vertical spring, stiffness $K$. Also attached to the mass is a viscous damper of rate $f$. The mass of the spring and the damper is negligible. The upper end of the spring may be displaced vertically. Derive the operational relation between displacements of the mass $x_2$ and of the upper end of the spring $x_1$.

$$\frac{x_2}{x_1} = \frac{K}{MD^2 + fD + K}$$

Determine the *amplitude* of the oscillations of $x_2$ when $x_1$ is moved sinusoidally at frequency $4\frac{1}{2}$ rad/s with an amplitude of $2\frac{1}{2}$ cm if the following data apply: $M = 12$ kg (26.46 lb), $K = 3$ N/cm (1.71 lb/in), $f = 0.4$ N/(cm/s) (0.288 lb/(in/s)).

(Approx. 4 cm. Note. $\omega_n = 5$, $\zeta = \frac{1}{3}$)

6 The pressure of a liquid in a pipeline is varying sinusoidally between 100 and 200 bar at a frequency of 10 Hz ($20\pi$ rad/s). Calculate the upper

and lower pressure readings which would be given by a Bourdon gauge connected to the pipeline if the gauge may be treated as a first order system with time constant $\frac{1}{5}$ s. Also calculate the limits (in bars) of the signal from another type of pressure-measuring device (for example strain gauge transducer, plus recorder), which may be treated as a second-order system with natural frequency 600 Hz, damping ratio 0.9.

($150 \pm (7.9/2)$ bar, lag $85\frac{1}{2}^{\circ}$; limits are $100 + 0.008$ and $200 - 0.008$ bar, lag $1.7^{\circ}$. Hence latter device gives precise readings.)

# 3 Hydraulic Frequency

This chapter is concerned with the primary source of oscillations in hydraulic systems, namely the interaction between the inertia of the moving parts and the compressibility of the fluid.

This source of oscillation is basic to all hydraulic devices and is of vital importance in design calculations. Other sources of oscillation can exist usually associated with detailed features of the components which go to make up a complete system. They may be related to the backlash in some linkage, pressure waves in pipelines or pumping pulsations for example but such subsidiary effects are not dealt with here and attention is confined to the inertia/compressibility effect.

Fluid trapped by a piston in a cylinder is compressible and acts like a spring. The piston (together with its rod and any load rigidly connected to it) will therefore behave as though it were spring mounted. Knowing the compressibility of the fluid and its volume together with the total mass of the moving parts, the natural frequency of such an oscillatory system may be calculated.

Three representative cases will be considered; a weight supported by a single-acting cylinder (figure 3.1), a double-acting cylinder with trapped fluid at each end of the cylinder (figure 3.2) and a double-acting cylinder with a long exhaust pipeline (figure 3.3). Some combination of these cases may have to be considered when dealing with real systems. Fluid may also be trapped by a hydraulic motor; when the motor shaft (and any inertia load attached to or geared to it) will behave as though it were mounted with torsional springs. One case will be considered (figure 3.4).

Fluid compressibility $\sigma$ is defined as the change in volume per unit volume for unit change in pressure. Bulk modulus $\beta$ is the reciprocal of compressibility and many hydraulic oils have a bulk modulus of about $17 \times 10^8$ N/m$^2$ (246 600 lb/in$^2$). Bulk modulus is drastically reduced by the presence of free air (in the form of bubbles, for example) in the fluid. Effective bulk modulus is also reduced by dilations of the container(s) in which the fluid is trapped.

Hydraulic frequency calculations are usually based on an assumption that all pressure variations are continuous with no discontinuities such as would occur with cavitation.

## 3.1 A Single-acting Hydraulic Jack

Consider a mass supported on a single-acting hydraulic jack as illustrated in figure 3.1.

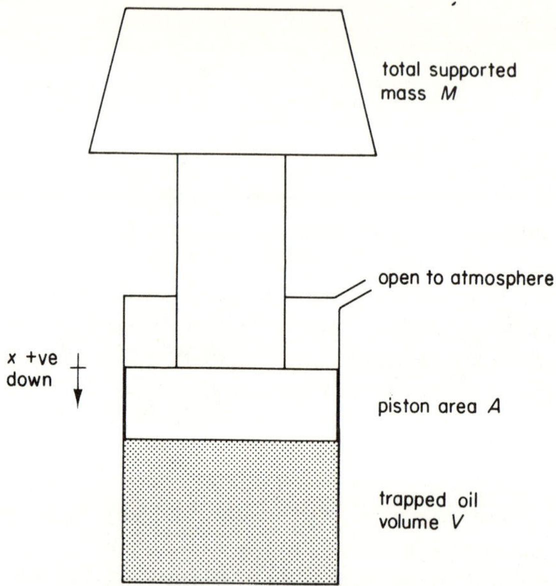


Figure 3.1 Single-acting cylinder with mass

Under static conditions, the pressure of the trapped oil (assuming no friction and no leakage) equals the total supported weight divided by the piston area. Small downward displacements cause increases in this pressure and small upward displacements cause decreases. A change in pressure from the static value will be designated $P$ and a displacement from the static position, $x$.

This change of pressure $P$ (assuming no leakage) will be directly proportional to the displacement $x$, according to the definition of bulk modulus or

$$P = \frac{\beta}{V} Ax$$

($x$ and hence $P$ may be $+$ ve or $-$ ve).

Under dynamic conditions (with no friction) the acceleration of the piston will be proportional to $P$ or

$$PA = - M \frac{\mathrm{d}^2 x}{\mathrm{d}t^2}$$

Hence the system equation in $x$ or $P$ represents a simple oscillatory system, that is

$$x = - \frac{V}{\beta} \frac{M}{A^2} \frac{\mathrm{d}^2 x}{\mathrm{d}t^2}$$

or

$$P = -\frac{V}{\beta}\frac{M}{A^2}\frac{\mathrm{d}^2 P}{\mathrm{d}t^2}$$

and the 'hydraulic frequency' $\omega_h$ is given by

$$\omega_h = \left(\frac{\beta}{V}\frac{A^2}{M}\right)^{1/2} \tag{3.1}$$

for a single-acting cylinder.

## 3.2  A Double-acting Cylinder

Consider a double-acting cylinder for which the total moving mass is $M$.

Referring to figure 3.2, there are two trapped volumes of fluid. The initial or static pressure is assumed to be the same on each side; $P_1$ and $P_2$ represent changes from this static pressure. Displacements (assumed to be small) from the initial position are termed $x$.

With no friction, the acceleration is proportional to the pressure difference, that is

$$(P_1 - P_2)A = -M\frac{\mathrm{d}^2 x}{\mathrm{d}t^2}$$

And for a displacement $x$ we have (with no leakage)

$$P_1 = \frac{\beta}{V_1}Ax \text{ and } P_2 = -\frac{\beta}{V_2}Ax$$

or

$$P_1 - P_2 = \beta A\left(\frac{1}{V_1} + \frac{1}{V_2}\right)x$$

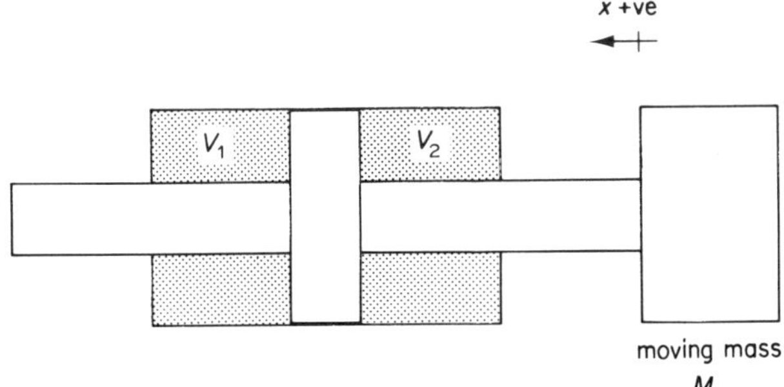


Figure 3.2  Double-acting cylinder with mass

whence the equation for the oscillatory system is

$$x = -\frac{1}{\beta(1/V_1 + 1/V_2)}\frac{M}{A^2}\frac{d^2x}{dt^2}$$

The hydraulic frequency is lowest when $V_1 = V_2$ (compared with any position of the piston where $V_1 \neq V_2$) and is of special interest. Usually the two volumes are equal with the piston in the midposition. With $V_1 = V_2$, the equation in $x$ reduces to

$$x = -\frac{V}{2\beta}\frac{M}{A^2}$$

and the 'hydraulic frequency' $\omega_h$ is given by

$$\omega_h = \left(\frac{2\beta}{V}\frac{A^2}{M}\right)^{1/2} \tag{3.2}$$

(where $V$ represents each half-cylinder volume) for a double-acting piston and it is sometimes written

$$\omega_h = \left(\frac{4\beta}{V_t}\frac{A^2}{M}\right)^{1/2} \tag{3.2a}$$

where $V_t$ represents the total cylinder volume.

As a numerical example of applying equation 3.2a consider a total mass of 2 tonnes (4 410 lb) with a piston of diameter 50 mm (1.97 in) and a rod of diameter 25 mm (0.98 in). Taking the effective bulk modulus of the fluid to be $12.4 \times 10^8$ N/m$^2$ (179 800 lb/in$^2$) and the total travel as 0.45 m (17.7 in)

$$\omega_h = \left(\frac{4 \times 12.4 \times 10^8 \times (0.001\ 47)^2}{0.000\ 661\ 5 \times 10^3 \times 2}\right)^{1/2} \approx 90 \text{ rad/s}$$

$$\approx 14\tfrac{1}{2} \text{ Hz}$$

(total volume 0.000 661 5 m$^3$) (or in British units

$$\omega_h = \left(\frac{4 \times 179\ 800 \times (2.278)^2 \times 386}{40.32 \times 4410}\right)^{1/2} = (8100)^{1/2}$$

area 2.278 in$^2$, volume 40.32 in$^3$).


## 3.3   A Double-acting Cylinder with a Long Exhaust Pipeline

An easily overlooked factor in estimating hydraulic frequencies is the mass of oil in pipelines. If the pipelines are of small diameter, then exceedingly high rates of fluid acceleration may be involved and, although the actual oil mass is small, its effect is not. This effect should be considered in every system and not only in the particular ram system considered here.

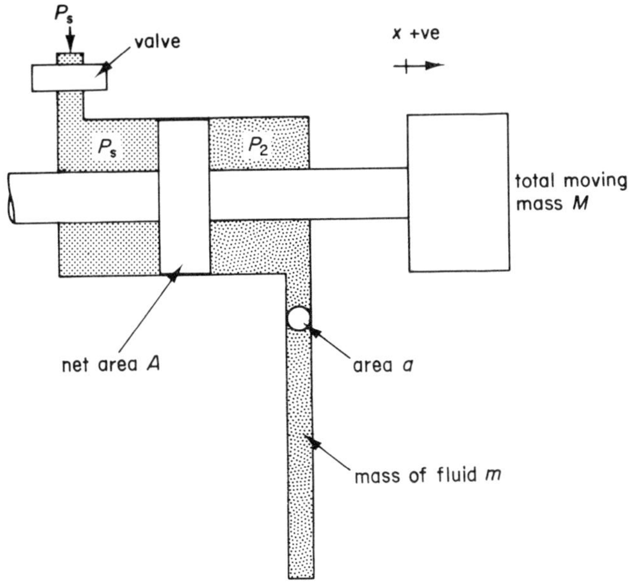


Figure 3.3

Consider a double-acting cylinder as illustrated in figure 3.3 supplied with fluid on the left and connected on the right by means of a pipeline to a tank at zero (atmospheric) pressure. We assume for clarity that there is no friction and no leakage. Let fluid be suddenly admitted at some constant pressure $P_s$ and let us predict the acceleration of the piston (a supply system with a large accumulator just before the left-hand port being suddenly connected by opening a valve, for example). With no friction or leakage, and no pressure in the right-hand chamber we should have

$$P_s A = M \frac{d^2 x}{dt^2} \tag{3.3}$$

(that is assuming the pressure in the right-hand chamber is atmospheric and that in the left $P_s$ above atmospheric). Equation 3.3 would give an overestimate of the acceleration because, during acceleration, the pressure must be positive at the pipeline entrance in order to accelerate the fluid contained in the pipe (this is $P_2$ which, with a steady flow, would also be slightly positive owing to fluid friction but here we neglect friction). Thus equation 3.3 is rewritten as

$$(P_s - P_2)A = M \frac{d^2 x}{dt^2} \tag{3.4}$$

and an approximate value for $P_2$ can be obtained (neglecting fluid friction) by assuming the oil in the pipe to move as a solid rod. For a pipeline of area $a$ containing a total mass $m$ of oil accelerating at rate $\mathrm{d}^2y/\mathrm{d}t^2$ we have

$$P_2 a = m\frac{\mathrm{d}^2y}{\mathrm{d}t^2}$$

The fluid in the pipeline moves and accelerates at $A/a$ times the rate of the fluid in the cylinder (neglecting fluid friction and compressibility effects) so

$$\frac{\mathrm{d}^2y}{\mathrm{d}t^2} = \frac{A}{a}\frac{\mathrm{d}^2x}{\mathrm{d}t^2}$$

Hence

$$P_2 a = m\,\frac{A}{a}\frac{\mathrm{d}^2x}{\mathrm{d}t^2} \quad \text{or} \quad P_2 = \frac{m}{a}\frac{A}{a}\frac{\mathrm{d}^2x}{\mathrm{d}t^2}$$

The term $P_2 A$ occurs in equation 3.4 and

$$P_2 A = \frac{m}{a}\frac{A}{a}\,A\,\frac{\mathrm{d}^2x}{\mathrm{d}t^2}$$

so, substituting in equation 3.4,

$$P_s A - m\,\frac{A^2}{a^2}\frac{\mathrm{d}^2x}{\mathrm{d}t^2} = M\frac{\mathrm{d}^2x}{\mathrm{d}t^2}$$

which for comparison with equation 3.3 is rewritten

$$P_s A = \left\{ M + \left(\frac{A}{a}\right)^2 m \right\}\frac{\mathrm{d}^2x}{\mathrm{d}t^2} \tag{3.5}$$

Compare equation 3.3;

$$P_s A = M\frac{\mathrm{d}^2x}{\mathrm{d}t^2}$$

The effective moving mass is $M + m(A/a)^2$ and $(A/a)^2$ may have an extremely large value. For example with a 75 mm (2.95 in) diameter piston and a 15 mm (0.59 in) pipeline—neglecting the rod area—$A/a$ equals $(75)^2/(15)^2$, that is 25 and $(A/a)^2 \approx 625$. Hence 0.2 kg (0.44 lb) could have the same effect as an additional mass fixed to the piston of 125 kg (275 lb). Hydraulic frequency calculations neglecting such additional mass can be grossly inaccurate and the above method enables its order of magnitude to be estimated.


## 3.4   An Oil Hydraulic Motor with Two Pipelines

Consider an oil hydraulic motor with two pipelines as indicated in figure 3.4.

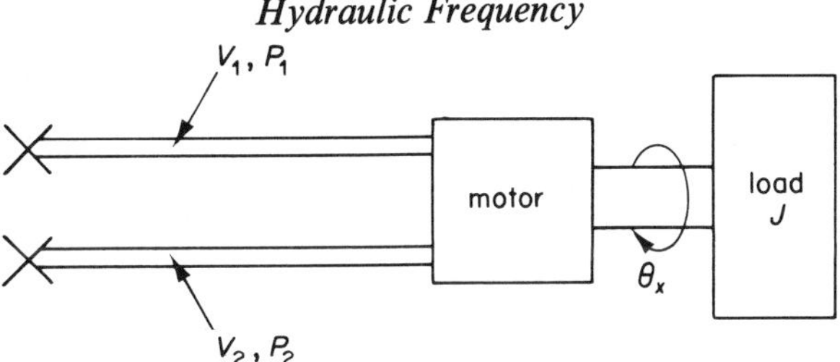



Figure 3.4 Motor with inertia load

If the two pipelines are closed, there are two volumes of oil, $V_1$ and $V_2$, which can be assumed to be trapped. There would be leakage within a real motor whenever the pressures in the two pipelines were different but here we shall neglect leakage completely. Also in a real motor, the two volumes would depend on the relative positions of the vanes, pistons or gears inside motor but here we shall assume that the volumes do not change significantly for small angular displacements of the motor shaft. Assuming, as before, that the two pressures are initially equal (and at some positive pressure level), it is clear that an angular displacement of the motor shaft would cause an increase of pressure in one pipeline and a decrease in the other. We also assume for simplicity that there is no fluctuation in flow with angular position—in effect therefore we are assuming that the motor has an infinite number of pistons, vanes or gear teeth so that the 'motor displacement' (designated $\delta_m$/rad) is a constant.

Consider that the shaft is forcibly moved to give an angular displacement of $\theta_x$ (radians). This would cause the motor to draw in a volume of fluid $\delta_m\theta_x$ on one side and deliver $\delta_m\theta_x$ on the other. These are two compressibility flows, increasing and decreasing $P_1$ and $P_2$ respectively. From the definition of bulk modulus

$$\delta_m\theta_x = \frac{V_1}{\beta}P_1 = -\frac{V_2}{\beta}P_2$$

or

$$P_1 - P_2 = \delta_m\beta\left(\frac{1}{V_1} + \frac{1}{V_2}\right)\theta_x$$

The torque which must be applied to the motor shaft is (with no friction)

$$\delta_m(P_1 - P_2)$$

Now, assuming the shaft were suddenly released, this torque would *cause* acceleration of the motor and its attached load in the absence of friction or other torques; hence

$$\delta_m(P_1 - P_2) = -J\frac{d^2\theta_x}{dt^2}$$

Thus the system equation in $\theta_x$ represents an oscillatory system

$$\delta_m \times \delta_m \beta \left( \frac{1}{V_1} + \frac{1}{V_2} \right) \theta_x = - J \frac{d^2 \theta_x}{dt^2}$$

and for most cases $V_1$ and $V_2$ are equal so that

$$\theta_x = - \frac{V}{2\beta} \frac{J}{\delta_m^2} \frac{d^2 \theta_x}{dt^2}$$

and the 'hydraulic frequency' $\omega_h$ is given by

$$\omega_h = \left( \frac{2\beta}{V} \frac{\delta_m^2}{J} \right)^{1/2} \tag{3.6}$$

(where $V$ represents the volume on each side of the motor) for a hydraulic motor or

$$\omega_h = \left( \frac{4\beta}{V_t} \frac{\delta_m^2}{J} \right)^{1/2} \tag{3.6a}$$

where $V_t$ represents the total trapped volume. Note that a motor may drive a lead screw to produce rectilinear movements of a table when $J$ would represent the inertia referred to the motor. Also note that the mass of fluid in connecting pipelines increases $J$.


**Problems**

1  A weight is supported by a hydraulic jack. Vertical oscillation of the weight occurs owing to the compressibility of the oil trapped under the piston of the jack. Estimate the frequency of the oscillations if the following data apply, assuming that the piston has negligible mass, that friction and leakage are also negligible and that the top of the piston is open to atmosphere: area of underside of piston 3 in$^2$ (0.001 935 m$^2$); volume of trapped oil 36 in$^3$ (0.000 59 m$^3$); weight being supported 2320 lb (1052.3 kg); effective bulk modulus of trapped oil 240 000 lb/in$^2$ or $16.55 \times 10^8$ N/m$^2$.

(15.9 Hz.)


2  Oil is supplied from a pumping unit to a fixed-displacement motor. Derive the (operational) relation between the *speed* of the motor and the flowrate from the pumping unit. Determine the maximum load inertia which can be coupled to the motor if the natural frequency of the system is to be limited to 25 Hz when the following data apply: motor

capacity 82 cm$^3$/rev; oil volume trapped between pump and motor 330 cm$^3$; effective bulk modulus $10^9$ N/m$^2$. Leakage and friction may be neglected but would be present in practice and would damp oscillations.

(About 1/50 kg m$^2$.)

# 4 Variable Pump Systems

This chapter makes use of the concepts already introduced to show how the dynamic characteristics of one type of complete hydraulic servo system can be predicted. The technique involves analysing the component parts separately and then combining the analyses to predict how the system will behave when the components are coupled together. In this chapter the only two components considered are a pump and a motor. As with all linearised analyses, the aim is to obtain a mathematical model which is readily dealt with but at the same time gives reasonably realistic predictions of actual system behaviour.

A variable-delivery pump is often used, when connected by two pipelines to a fixed-displacement motor, for controlling the speed of the motor or the angular position of the motor shaft by adjusting the flowrate from the pump. The pump is driven at constant speed and the flowrate adjusted by altering the angular position of a swash plate (in the case of axial piston pumps) or the eccentricity of a control ring (in the case of vane or segment pumps) with a control rod or lever protruding from the pump. Fully reversible flow is normally available, the flowrate being zero when the control lever or other device is in its central or null position. Both connecting pipelines are able to draw in fluid through non-return valves from a booster supply (which is provided to prevent very low pressures occurring and to make up for external leakages from pump and motor). Across the two pipelines a leakage path may be provided to improve dynamic behaviour (by increasing the damping of induced oscillations). Relief valves are incorporated to reduce the risk of excessive pressures. The general arrangement of such pump control systems is shown in figure 4.1.

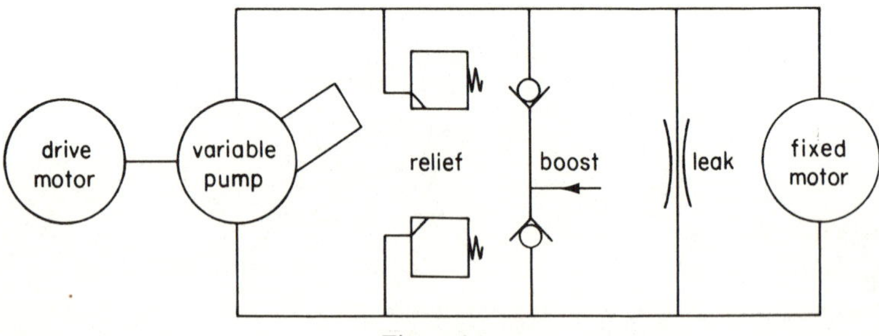


Figure 4.1

## 4.1 The Pump

More fluid enters a pump than is delivered at the outlet owing to *external* leakage. For many cases, however, this external leakage can be neglected and the 'pump flowrate' $q_p$ taken to mean the volume flowrate at the entry or at the exit. Now *internal* leakage causes this flowrate to be less than the volume swept out (by the pistons or vanes) per unit time. This internal leakage, past the pistons for example and across the porting of the pump, is normally taken as directly proportional to the difference in pressure between inlet and outlet (flow through the small clearances being taken as laminar). The pump flowrate for a fixed-displacement pump would be

$$q_p = \alpha_p - L_p(P_1 - P_2) \tag{4.1}$$

The pump flowrate for a variable-delivery pump is taken as

$$q_p = \imath\alpha_p - L_p(P_1 - P_2)$$

where $L_p$ is assumed constant, $P_1$ and $P_2$ are the outlet and inlet pressures respectively, $\alpha_p$ is the pump swept volume (per unit time) for the particular speed at which it happens to be running and $\imath$ is the proportional setting of the pump flow control. ($\imath$ may also be quoted as a physical displacement (measured in millimetres, for example) when $\alpha_p$ is then understood to represent the swept volume per unit displacement of the control ((litres/s)/mm, for example).)

Pump speed fluctuations are often neglected—particularly if the pump is driven by a motor or engine of large inertia.

## 4.2 The Motor

'Motor flowrate' $q_m$ is similarly taken to mean the volume flowrate at either the entry or the exit (external leakage being neglected) and is quoted as

$$q_m = \delta_m n + L_m(P_1 - P_2) \tag{4.2}$$

where $n$ is the (instantaneous) motor speed (rad s$^{-1}$), $\delta_m$ is the motor swept volume per radian (the motor capacity), $P_1$ and $P_2$ are the inlet and outlet pressures respectively and $L_m$ is assumed to be a constant.

The torque in the motor provided by the pressure difference $P_1 - P_2$ is ideally given by

$$\tau = \delta_m(P_1 - P_2) \tag{4.3}$$

(that is torque equals work done per radian by the fluid).

A smaller torque is available to accelerate the motor shaft and the things connected to it because of friction. This motor friction may be significant in size and complicated in form (including viscous, coulomb and pressure-dependent terms). For analytical purposes, friction may be either neglected or assumed to be purely viscous.

If friction is neglected, then with a flywheel or other pure inertia load connected to the motor shaft, the whole of the torque will be used to give acceleration or

$$\tau = J\,\frac{\mathrm{d}n}{\mathrm{d}t} \tag{4.4}$$

(where $J$ is the total inertia of the motor and load).

If the friction is assumed to be purely viscous and (taking load and motor together) of magnitude $f$ units (for example $(\mathrm{N\,m})/(\mathrm{rad/s})$) the torque equation would be

$$\tau = J\,\frac{\mathrm{d}n}{\mathrm{d}t} + fn \tag{4.4a}$$

The load being driven by the hydraulic motor could also be subject to other applied torques (or forces through a rack and pinion say) when the torque equation becomes

$$\tau = J\,\frac{\mathrm{d}n}{\mathrm{d}t} + fn + \tau_{\mathrm{app}} \tag{4.4b}$$

## 4.3   Open Loop Systems

If the pump control lever or rod is adjusted independently from the rest of the system, then the system is termed 'open loop'.

### 4.3.1   *Steady State Operation*

Under steady state conditions the motor speed would be proportional to the pump control setting (that is $n = \imath\alpha_{\mathrm{p}}/\delta_{\mathrm{m}}$) if there were zero load (frictional or otherwise) at the motor. This idealised (no load) case would not occur in practice; it would involve zero pressure difference between the two pipelines.

Under steady state conditions with pure viscous friction (of motor and load), the pressure difference would vary directly with the motor speed and equal $fn/\delta_{\mathrm{m}}$. With a total leakage coefficient $L$ for the system, the speed would be given by $n = \imath\alpha_{\mathrm{p}}/(\delta_{\mathrm{m}} + Lf/\delta_{\mathrm{m}})$.

The steady state speed would be affected by any torque applied externally to the motor. If the torque were sufficiently large, the motor would stall. Under stall conditions, the pressure difference would cause either all the pump flowrate to go in leakage or the relief valves to operate (that is at pressure $P = \tau_{\mathrm{app}}/\delta_{\mathrm{m}}$).

### 4.3.2 *Dynamic Analysis*

The transient characteristics of the open loop system would be influenced by the inertia (load plus moving parts of motor, total $J$) and by the volumes of fluid in the connecting pipelines between pump and motor. It is assumed that the pump speed is unchanging.

For dynamic analysis, it is also assumed (a) that a leakage path exists, through which the flowrate is proportional to the difference in pressure between the two pipelines, (b) that pressure losses along both pipelines are negligible and (c) that pressures in each pipeline vary continuously. Friction is completely neglected. A system diagram is shown in figure 4.2.

The flowrate from the pump $q_p$ goes to compress the oil volume $V_1$ and provides both the leakage flowrate and the motor flowrate or

$$q_p = \frac{V_1}{\beta}\frac{dP_1}{dt} + L_\ell(P_1 - P_2) + q_m \tag{4.5}$$

The flowrate $q_m$ from the motor goes to compress the oil volume $V_2$ and provides, when added to the leakage flow, the flow entering the pump or

$$q_m + L_\ell(P_1 - P_2) = \frac{V_2}{\beta}\frac{dP_2}{dt} + q_p \tag{4.6}$$

(It is of interest that both the pressures and their rates of change may be positive or negative. The analysis so far also leads to the prediction that $dP_1/dt = -dP_2/dt$ at all times if $V_1 = V_2$.)

Adding equations 4.5 and 4.6 (and dividing by 2), substituting $P$ for $P_1 - P_2$ and also taking the system to be symmetrical so that $V_1 = V_2 = V$, we obtain

$$q_p = \frac{V}{2\beta}\frac{dP}{dt} + L_\ell P + q_m \tag{4.7}$$

(note $V$ is the volume in one-half of the system).

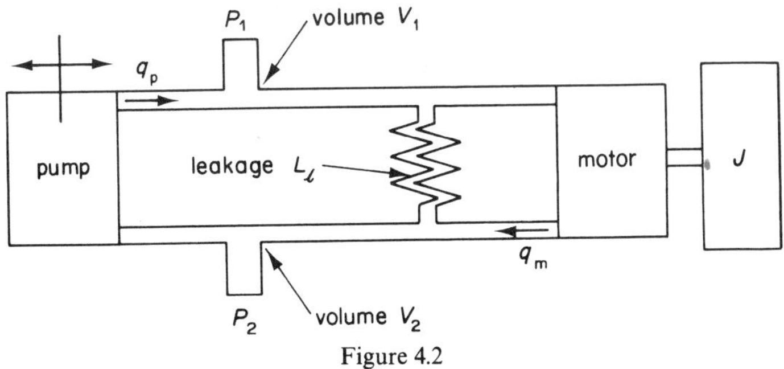


Figure 4.2

Substituting the pump equation 4.1a and the motor flowrate equation 4.2 into equation 4.7 gives

$$\imath\alpha_{\mathrm{p}} = \frac{V}{2\beta}\frac{\mathrm{d}P}{\mathrm{d}t} + (L_{\ell} + L_{\mathrm{p}} + L_{\mathrm{m}})P + \delta_{\mathrm{m}}n \qquad (4.8)$$

The motor torque equation, 4.3, is equated to the frictionless torque equation, 4.4, in order to eliminate $P$ from equation 4.8

$$P = \frac{J}{\delta_{\mathrm{m}}}\frac{\mathrm{d}n}{\mathrm{d}t}$$

hence

$$\frac{\mathrm{d}P}{\mathrm{d}t} = \frac{J}{\delta_{\mathrm{m}}}\frac{\mathrm{d}^2 n}{\mathrm{d}t^2}$$

This gives

$$\imath\alpha_{\mathrm{p}} = \frac{V}{2\beta}\frac{J}{\delta_{\mathrm{m}}}\mathrm{D}^2 n + L\frac{J}{\delta_{\mathrm{m}}}\mathrm{D}n + \delta_{\mathrm{m}}n \qquad (4.9)$$

(where $\mathrm{D} \equiv \mathrm{d}/\mathrm{d}t$ and $L = L_{\ell} + L_{\mathrm{p}} + L_{\mathrm{m}}$, the total leakage coefficient), so that the operational relation between the (output) motor speed and the (input) pump control displacement is

$$\frac{n}{\imath} = \frac{\alpha_{\mathrm{p}}/\delta_{\mathrm{m}}}{(1/\omega_{\mathrm{h}}^2)\mathrm{D}^2 + (2\zeta/\omega_{\mathrm{h}})\mathrm{D} + 1} \qquad (4.10)$$

where

$$\frac{1}{\omega_{\mathrm{h}}^2} = \frac{V}{2\beta}\frac{J}{\delta_{\mathrm{m}}^2} \quad \text{and} \quad \frac{2\zeta}{\omega_{\mathrm{h}}} = \frac{LJ}{\delta_{\mathrm{m}}^2}$$

(for a frictionless motor).

Equation 4.10 indicates that any displacement of the pump control (to cause changes of the pump flowrate) will be followed by oscillation of the motor speed. The natural frequency of the oscillations will be $\omega_{\mathrm{h}}$, and their rate of decay will be governed by $\zeta$ (with $\zeta$ determined, in the frictionless case, by the total leakage coefficient).

Equations similar to equation 4.10 may be derived for other motor load conditions, for example by equating 4.3 to 4.4a or 4.4b before substituting in equation 4.8. For the viscous friction case (equation 4.4a), it will be found that the damping factor $\zeta$ increases.


## 4.4   Closed Loop (Position Control) Systems

The pump flow control (lever) may be operated automatically to 'close the loop'. Consider a system devised to control automatically the angular position of the motor shaft. Some input device is assumed to exist (on which is selected the desired angular position $\theta_i$) plus some arrangement to displace the pump control lever by an amount which is proportional to the difference between the input or demand position $\theta_i$ and the output (motor shaft position $\theta_o$). A purely mechanical system can be devised. In the analysis, we assume that pump control lever displacements are *directly* proportional to this difference (that is the 'error') although more complicated cases can be similarly dealt with. The assumption may be written

$$\imath = k(\theta_i - \theta_o)$$

The dynamic characteristics of this closed loop system may be investigated using the equations already developed. Being concerned with position rather than speed, it is convenient to write $D\theta_o$ for $n$ and for the *frictionless* case, equation 4.9 may be rewritten

$$\alpha\imath = \frac{V}{2\beta}\frac{J}{\delta_m}D^3\theta_o + L\frac{J}{\delta_m}D^2\theta_o + \delta_m D\theta_o \qquad (4.11)$$

Substituting $\alpha_p\imath = \alpha_p k(\theta_i - \theta_o)$ gives

$$\alpha_p k\theta_i = \frac{V}{2\beta}\frac{J}{\delta_m}D^3\theta_o + L\frac{J}{\delta_m}D^2\theta_o + \delta_m D\theta_o + \alpha_p k\theta_o$$

or

$$\theta_i = \frac{V}{2\beta}\frac{1}{\alpha_p k}\frac{J}{\delta_m}D^3\theta_o + \frac{L}{\alpha_p k}\frac{J}{\delta_m}D^2\theta_o + \frac{\delta_m}{\alpha_p k}D\theta_o + \theta_o \qquad (4.12)$$

so that the operational relation between output and input is in the form

$$\frac{\theta_o}{\theta_i} = \frac{1}{a_0 D^3 + a_1 D^2 + a_2 D + a_3}$$

a third-order linear differential equation with constant coefficients in which

$$a_0 = \frac{V}{2\beta}\frac{J}{\delta_m{}^2}\frac{\delta_m}{k\alpha_p} = \frac{1}{\omega_h{}^2}\frac{\delta_m}{\alpha_p}\frac{1}{k}$$

$$a_1 = L\frac{J}{\delta_m{}^2}\frac{\delta_m}{k\alpha_p} = \frac{2\zeta}{\omega_h}\frac{\delta_m}{\alpha_p}\frac{1}{k}$$

$$a_3 = 1, \qquad a_2 = \frac{\delta_m}{\alpha_p}\frac{1}{k}$$

(see equation 4.10 for $\omega_h$ and $\zeta$). The four system parameters which occur are as follows: $\omega_h$, the natural frequency referring to the motor, its inertia load and the trapped oil volume; $\zeta$, the damping factor referring to the leakage coefficient in particular; $\delta_m/\alpha_p$, the ratio of motor displacement (say l/rad) to the pump swept volume (say l/s); $k$ representing the gain (lever ratios, etc.) of the feedback linkage arrangement.

## 4.5 Practical Systems

A major weakness in the analysis is that both pressures are assumed to vary continuously whereas in fact discontinuities must occur whenever one of the pressures falls to the booster pressure where it will remain constant until it next begins to increase. Now the analysis involves the prediction that at all instants the two pressures, $P_1$ and $P_2$, will be equal but opposite in sense because their initial values are the same and their rates of change are equal but opposite. A simple modification is to assume that the pressure difference $P$ will always have half its predicted value. A convenient way of allowing for this is to use $2V$ (to replace $V$) in equations 4.10 and 4.12.

An investigation of hydrostatic transmissions made at Bath University is reported in Bowns and Worton-Griffiths (1972).

## Problem

1 A variable pump fixed-displacement motor type of control (see figure 4.2) is used in closed loop form to position a rotary turret of mass 50 kg and radius of gyration 0.666 m (inertia $J = 22.18$ kg m$^2$ or 75 785 lb in$^2$). The motor has capacity $\delta_m$ of 0.08 l/rad (4.88 in$^3$/rad). The pump with its control and feedback linkage are such that the flowrate from the pump is at the rate $k\alpha_p$ of 0.396 l/s (24.17 in$^3$/s) for every radian error (that is for every radian difference between a selected demand or input position and the actual angular position of the turret at any instant). The volume of oil trapped in *each* line between pump and motor is 0.5 l (30.51 in$^3$) and the effective bulk modulus of the oil is $14 \times 10^8$ N/m$^2$ (215 350 lb/in$^2$). Total internal leakage (within motor, pump and ancillaries) is proportional to the pressure difference across the pump or motor at the rate of $1.46 \times 10^{-12}$ (m$^3$/s)/(N/m$^2$) (0.0006 (in$^3$/s)/(lb/in$^2$)). Friction may be neglected. Show that the closed loop (operator) relation between the output $\theta_o$ and the input $\theta_i$ can be written

$$\left(\frac{1}{8000}\,D^3 + \frac{1}{976}\,D^2 + \frac{1}{4.95}\,D + 1\right)\theta_o = \theta_i$$

which reduces to

$$\left(1 + \frac{1}{5}\,D\right)\left\{\frac{1}{(40)^2}\,D^2 + \frac{2 \times 0.04}{40}\,D + 1\right\}\theta_o = \theta_i$$

noting that $\omega_h = 40.2$ and $\zeta = 0.1$.

# 5 Linear Control Theory

This chapter outlines ways of extending the methods of dynamic analysis used in chapter 2 in order to cope with the characteristics of complete servo systems. It begins with an indication of the behaviour to be expected from systems governed by third-order equations (such as that analysed in chapter 4). Then it deals with some generalisations and techniques involving block diagrams and vectorial representations of harmonic response. These lead up to a summary of the criteria which can be used to assess the stability of servo systems in general and of hydraulic servomechanisms in particular.

## 5.1  Algebraic Stability Criterion (Routh–Hurwitz)

First-order and second-order systems have governing equations with positive coefficients. First-order equations have the form $1 + T\lambda = 0$ and $T$ is positive. Second-order equations have the form $a_0\lambda^2 + a_1\lambda + a_2 = 0$ where $a_0 = 1$, $a_1 = 2\zeta\omega_n$ and $a_2 = \omega_n{}^2$, each coefficient $a_0$, $a_1$ and $a_2$ being positive. Transient response is characterised by the root(s) of the subsidiary equation, $-1/T$ for first order, $-\omega_n\zeta \pm i\omega_n(1 - \zeta^2)^{1/2}$ for second order, the real part in each case being negative. With first-order systems the term $\exp(-t/T)$ (for example in equations 2.6 and 2.7) and with second-order systems the term $\exp(-\zeta\omega_n t)$ (for example in equations 2.10 and 2.11) indicate that, with these two types of systems, any transients will decay.

Higher-order systems are more common. One example is the variable pump position control system described in chapter 4 which can be treated as a third-order system.

Third-order systems have subsidiary equations of the form $a_0\lambda^3 + a_1\lambda^2 + a_2\lambda + a_3 = 0$. The step response of such systems can be considered as a second-order (oscillatory) response superimposed on a first-order (exponential) response as illustrated in figure 5.1, with the difference that transients do not necessarily decay. As with simpler systems, the roots of the subsidiary equation determine response and these roots must have negative real parts if the system is to be stable. It is not sufficient for all the coefficients to be positive because the real part of one (or more) roots might still be positive. For example, a system with the subsidiary equation $\lambda^3 + \lambda^2 + 4\lambda + 30 = 0$ has roots of $-3$ and $1 \pm 3i$ (the equation factorises as $(\lambda + 3)(\lambda - 1 + 3i)(\lambda - 1 - 3i)$) and transients are characterised by equations of the form $Ae^{-3t} + Be^{(1+3i)t} + Ce^{(1-3i)t}$, the last two terms representing an oscillation of *increasing* amplitude similar to that

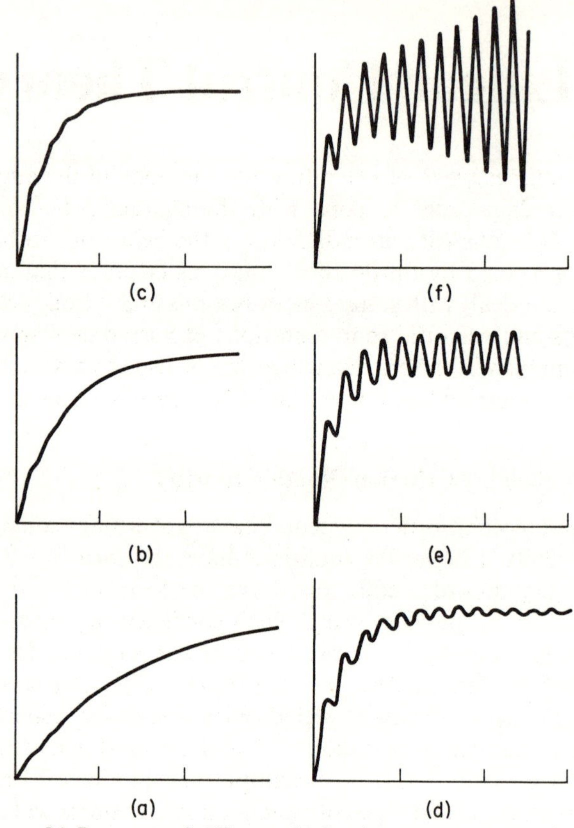

Figure 5.1 Response of different third-order systems to step inputs. For time bases (abscissae) scaled in seconds the plots would refer to systems with subsidiary equations of the form (cf. equation 4.13)

$$\frac{1}{\omega_g{}^2}\lambda^3 + \frac{2\zeta}{\omega_g}\lambda^2 + \lambda + K = 0$$

where in all cases $\omega_g = 27$ rad s$^{-1}$; $\zeta \approx 0.18$ for (a), (b) and (c), $\zeta \approx 0.07$ for (d), (e) and (f); $K$ equals about 1.0 for (a), 2.1 for (b), 3.2 for (c) and (d), 4.3 for (e) and 5.3 for (f)

shown in the top right-hand diagram of figure 5.1. The other requirement (in addition to all coefficients being positive) relates to the numerical values of the coefficients: $a_1 a_2 > a_3 a_0$ is necessary for stability of third-order systems; $(a_1 a_2 a_3 - a_1 a_4 a_1 - a_3 a_0 a_3 > 0$ is necessary for stability of fourth-order systems). For description and details of Routh or Routh–Hurwitz stability criterion, see Brown (1965).

## 5.2 Open Loop Relations

### 5.2.1 *First-order Example*

The hydraulic position control system illustrated in figure 2.3 can under certain conditions be represented as a first-order system and in block diagram form as figure 5.2. The block on the right indicates that $q = AD\theta_o$. The difference between input and output is halved to give the valve displacement or $x = \frac{1}{2}(\theta_i - \theta_o)$ and $k$ is the flow coefficient for the valve or $q = kx$. The output–error (open loop) relation for the system is $AD\theta_o = (k/2)\theta$, or

$$\frac{\theta_o}{\theta} = \frac{k}{2AD} \text{ written } K \text{ fn(D)}$$

$K$ fn(D) is the *open loop* (operator) relation between the output and the error.

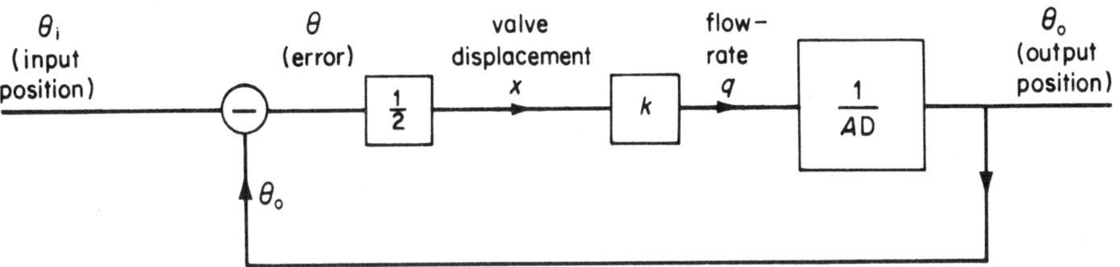


Figure 5.2 Block diagram for first-order approximation of the hydraulic system in figure 2.3

The error $\theta$ is defined as the difference between the input and the output, that is $\theta = \theta_i - \theta_o$ and the connection between the output–error relation, $\theta_o/\theta$, and the output–input relation $\theta_o/\theta_i$ is conveniently expressed as

$$\frac{\theta_o}{\theta_i} = \frac{K\text{fn}(D)}{1 + K\text{fn}(D)}$$

because

$$\frac{\theta_i}{\theta_o} = \frac{\theta_o + \theta}{\theta_o} = 1 + \frac{1}{K\text{fn}(D)} = \frac{K\text{fn}(D) + 1}{K\text{fn}(D)}$$

In this example

$$\frac{\theta_o}{\theta_i} = \frac{k/2AD}{1 + k/2AD} = \frac{1}{1 + (2A/k)D}$$

the same relation obtained in chapter 2.

### 5.2.2 *Third-order Example*

The third-order variable pump position control system considered in

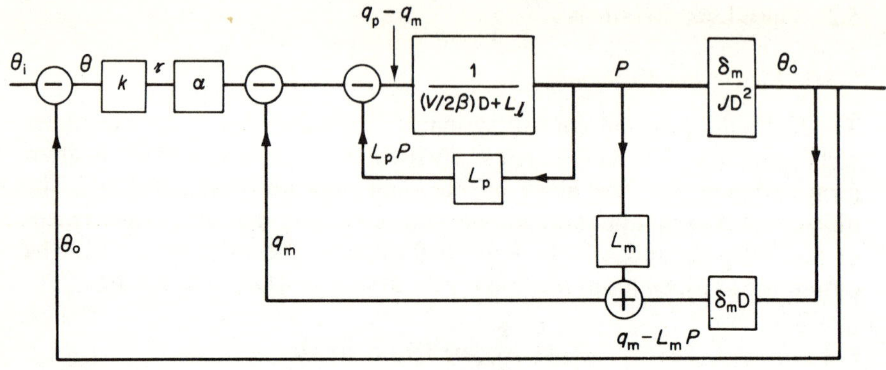



(a)

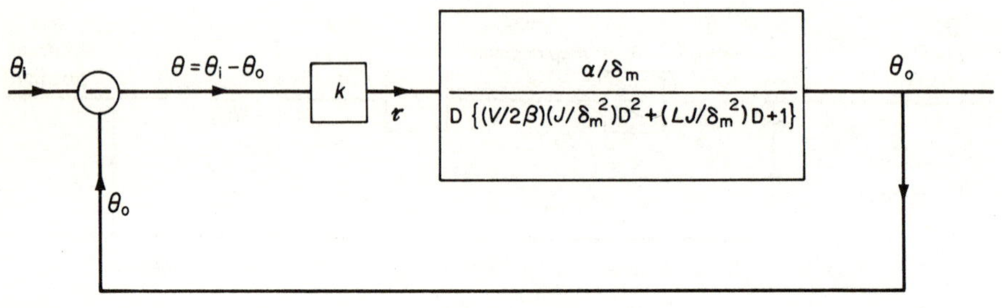



(b)

Figure 5.3 Block diagrams for third-order variable pump control systems of figures 4.1 and 4.2

chapter 4 may also be represented in block diagram form (note that causes and effects are unidirectional as indicated by arrows). The relations for the system, equations 4.1, 4.2, 4.3 and 4.7, are represented in figure 5.3(a). However, in this case, the diagram is simpler when based on equation 4.11 as shown in figure 5.3(b).

The output–error (open loop) relation for the system is

$$\frac{\theta_o}{\theta} = \frac{k\alpha/\delta_m}{D\{(V/2\beta)(J/\delta_m^2)D^2 + (LJ/\delta_m^2)D + 1\}} \quad \text{written } K\,\text{fn}(D)$$

and, as before, the connection with the (closed loop) relation is

$$\frac{\theta_o}{\theta_i} = \frac{K\text{fn}(D)}{1 + K\text{fn}(D)}$$

which when substituted gives equation 4.12.

### 5.2.3  *General Case*

Error-actuated feedback systems (with corrective action determined by the error) have an open loop relation

$$\frac{\theta_o}{\theta} = K\text{fn}(D)$$

and closed loop relation

$$\frac{\theta_o}{\theta_i} = \frac{K\text{fn}(D)}{1 + K\text{fn}(D)}$$

and a block diagram of the type shown in figure 5.4.

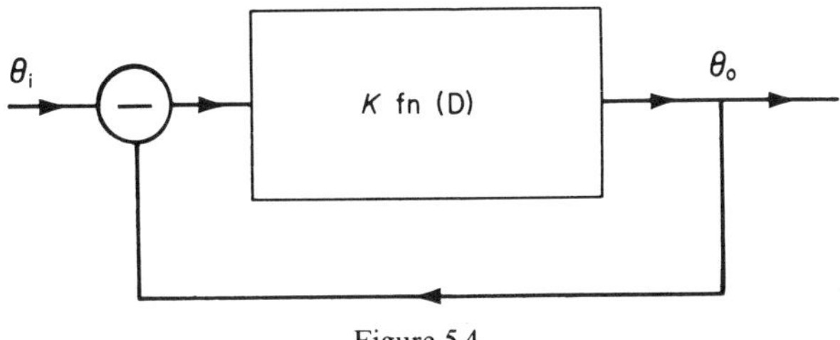


Figure 5.4

### 5.2.4  *Harmonic Input*

It was illustrated in chapter 2, for the special case of simple harmonics, that input–output relations were obtainable for linear systems by replacing D in the appropriate operational relation by $i\omega$ and

$$\frac{\theta_o}{\theta} = K\text{fn}(i\omega) \text{ for } \theta = \exp(i\omega t)$$

$$\frac{\theta_o}{\theta_i} = \frac{K\text{fn}(i\omega)}{1 + K\text{fn}(i\omega)} \text{ for } \theta_i = \exp(i\omega t)$$

### 5.2.5  *Open Loop Harmonic Response  Locus*

For simple harmonic inputs to error-actuated systems (with $\theta = \theta_i - \theta_o$), the vectorial representations can be extended to include all three vectors. In a vectorial representation for a complete closed loop system of $\theta_o$ and $\theta_i$ with $\theta_i = \exp(i\omega_a t)$ a diagram such as figure 5.5(a) occurs with vectorial subtraction giving the error vector $\theta$.

Retaining the same relative magnitudes, figure 5.5(a) may be redrawn as figure 5.5(b) with the error vector $\theta$ as the unit vector (representing $\exp(i\omega_a t)$.)

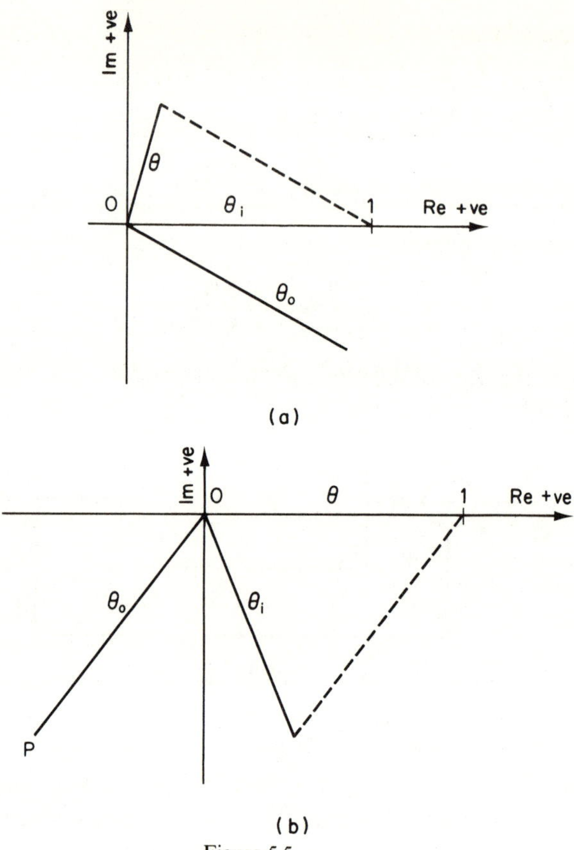



Figure 5.5

The vector $\theta_o$ of figure 5.5(b) represents the oscillation of the output when the error is forcibly varied sinusoidally and hence represents the open loop response of the servo system.

### 5.2.6　*Open Loop Testing*

To perform an open loop harmonic response test on the hydraulic servo sketched in figure 2.3 (see block diagram, figure 5.2), the feedback lever would be first disconnected at the piston rod and a pin fixed to constrain the upper end of the lever from moving horizontally. The input would then be moved back and forth sinusoidally at different frequencies (with some arbitrarily defined 'unit' amplitude). Input and output would be monitored to measure the amplitude and phase lag of the piston. At each test frequency, one output vector—such as 0P in figure 5.5(b)—would be obtained.

Similarly for the variable pump position control system (figures 4.1 and 5.3), any feedback connection would be removed and the pump control lever would be moved back and forth sinusoidally. Measurements of the resulting angular movements of the motor shaft would reveal the phase lag and, taking into account the (mechanical) ratio $k$, the amplitude ratio, hence defining an output vector for each frequency.

## 5.3 Nyquist Stability Criterion

The behaviour of any linear system may be represented in the form of its *open loop* harmonic response locus (the line which joins the end points of all the vectors such as 0P in figure 5.5(b) for all frequencies between 0 and $\infty$). Each individual vector represents the oscillation of the output when the error is forced at a particular frequency with unit amplitude (the feedback loop being disconnected). Point P is located on the diagram for a particular frequency $\omega_a$ as the complex number $K\,\mathrm{fn}(i\omega_a)$.

It will usually be found that at one particular frequency (say $\omega_x$) the output is exactly out of phase with the error (that is the open loop phase lag is 180°) when the particular vector 0P must lie along the negative real axis. The length of this (180°) vector is of crucial importance to the stability of a system.

Consider a notional system for which the 180° vector has a length greater than unity (say 1.1). This would imply that, for a particular frequency $\omega_x$, the term $K\mathrm{fn}(i\omega_x) = -1.1$. Refer to the general block diagram of figure 5.4 in order to visualise the effect of such a result. Assume that there is no input ($\theta_i$ can be ignored) but assume that some slight induced oscillation exists in the form of an error sine wave (that is $\theta$ is a sine wave of frequency $\omega_x$ and of small amplitude). The output $\theta_o$ would then be a sine wave slightly larger in amplitude than $\theta$ and 180° out of phase with it. Owing to the inherent nature of negative feedback systems, the negative of the output becomes the error when there is no input and the sine wave $-\theta_o$ is exactly in phase with the sine wave $\theta$ which caused it. In this case, however, it is larger in amplitude. The net result would inevitably be oscillations of ever-increasing amplitude. The system would be *unstable.*

As a matter of interest were the length of the vector exactly equal to 1, the result would be an oscillation which simply continued indefinitely (marginal stability).

The *Nyquist* stability criterion, in its simplest form, is that the *open loop* harmonic response locus must *not* enclose the $-1$ point. (See West (1953) for description and Brown (1965) for mathematical derivation of Nyquist criterion.)

## 5.4 Adequate Stability

### 5.4.1 *Gain and Phase Margins*

Any closed loop system with an open loop locus which actually crossed the negative real axis at the $-1$ point would oscillate continuously and the

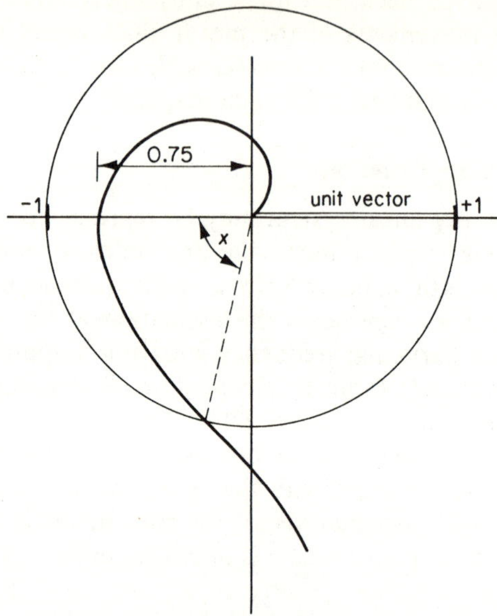



Figure 5.6 Nyquist diagram: phase margin $= x$; gain margin $= 1/0.75$ or 2.5 dB

proximity of the open loop locus to this point is some measure of the oscillatory tendencies of a system. Two empirical measures are used for assessing adequate stability of a control system. These are the 'phase margin' and the 'gain margin'. The phase margin is the number of degrees by which the open loop phase angle is less than 180° when the amplitude ratio is unity. The gain margin is associated with the amount by which the open loop amplitude ratio is less than unity when the open loop phase angle is 180° (lagging). The gain margin is usually measured in decibels and, if the length of the 180° vector were 0.75 for example, then the gain margin would equal $20 \log_{10} 0.75$ or 2.5 dB. An open loop locus for a stable system is shown in figure 5.6.

### 5.4.2  *Logarithmic Locus*

Logarithmic scale diagrams are used for open loop data when they are termed *Bode diagrams*. A system is stable if the amplitude curve is below the 0 dB line (amplitude ratio $< 1$) when the phase angle is 180°. A representative Bode diagram for a stable system, with gain and phase margins marked, is given in figure 5.7.

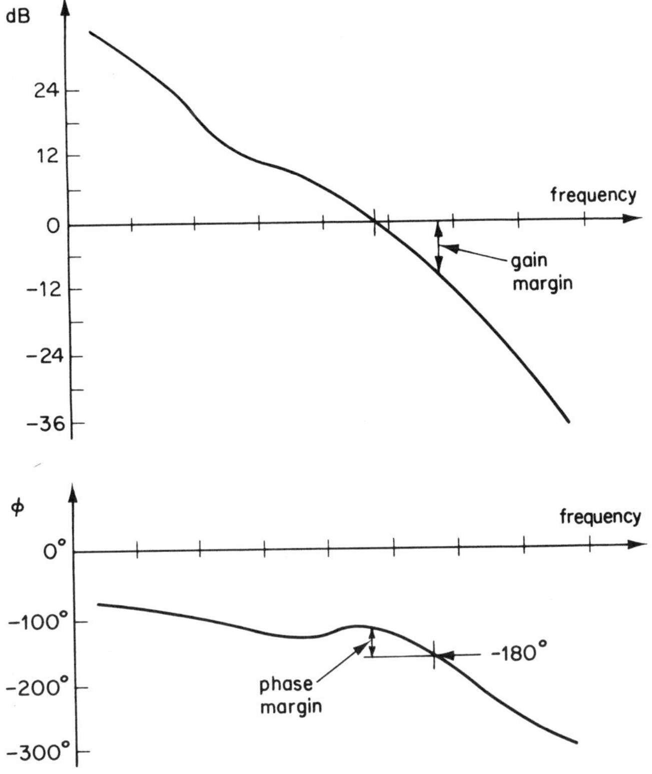


Figure 5.7 Example of Bode diagram

### 5.4.3 *Maximum Closed Loop Dynamic Magnification*

In view of the relationship between input, output and error vectors (see figure 5.5) it is possible to relate closed loop harmonic response to open loop response by graphical methods. In particular, the maximum closed loop dynamic magnification $\mathcal{M}_{max}$ can be obtained from the open loop harmonic response locus. A curve can be superimposed on the open loop locus which defines a particular ratio between the lengths of the $\theta_o$ and $\theta_i$ vectors—this ratio is $\mathcal{M}$ (the closed loop dynamic magnification). Such curves are coaxial circles which may be superimposed on polar plots. Orthogonal curves define closed loop phase angles. An open loop locus will, for some particular frequency, be tangential to one of these circles and the $\mathcal{M}$ value of this circle will be $\mathcal{M}_{max}$ for the closed loop system, see figure 5.8.

On logarithmic axes, but in this case with the open loop amplitude ratio (in dB) plotted directly against open loop phase angle, similar sets of

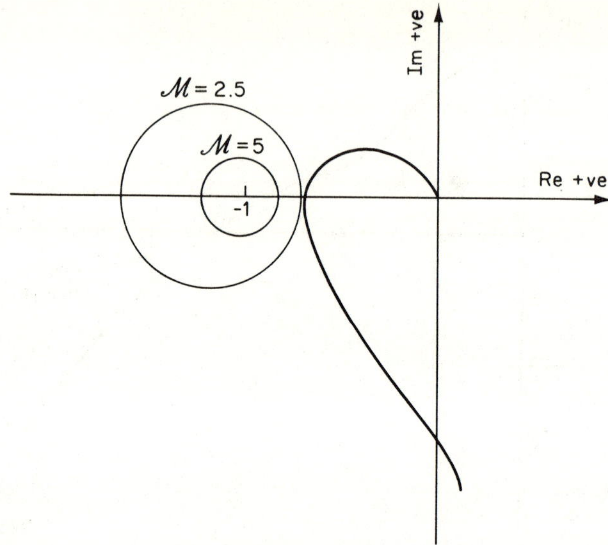


Figure 5.8  Hall diagram

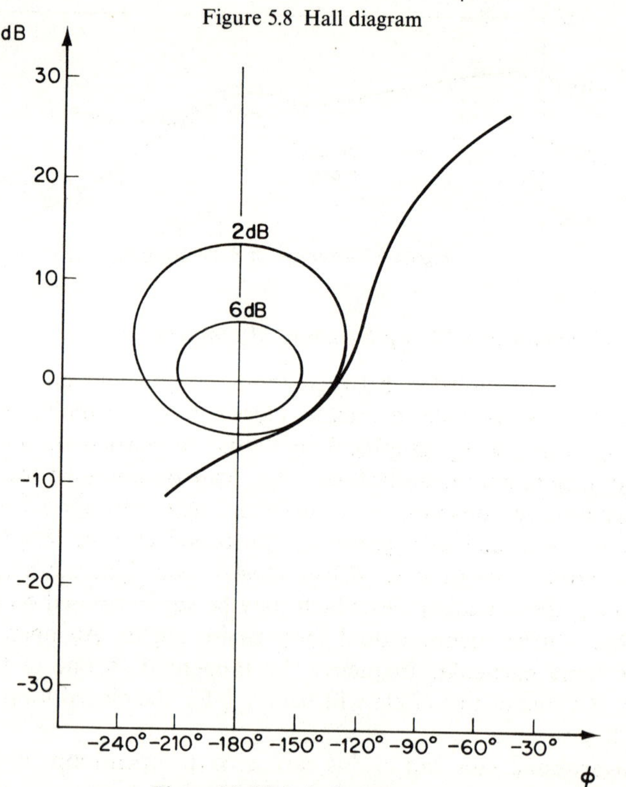


Figure 5.9  Nichols diagram

curves are obtained. The tangent point between one of these curves and the open loop locus defines $\mathcal{M}_{\text{max}}$ for the closed loop system, see figure 5.9.

## Problem

1 Estimate the phase margin and the gain margin of a system which has the following open loop (operator) relation given that $K = 2\mathrm{s}^{-1}$; the natural frequency $\omega_n = 8$ rad/s; and the damping factor $\zeta = 0.2$.

$$\frac{\theta_o}{\theta} = \frac{K}{D\{(1/\omega_n{}^2)D^2 + (2\zeta/\omega_n)D + 1\}}$$

(It is convenient to allow sufficient space on a Bode diagram for about 8 octaves, 80 decibels and $270°$ phase lag.) Note that the techniques relating to figures 8.6 and 8.7 may be applied or the complex numbers $\theta_o/\theta$ with D replaced by $i\omega$ may be graphed (for example as vectors) using $\omega = 2, 4, 8, 16$, etc.

(4 dB and approx. $90°$.)

# 6 Pumps

This chapter represents a respite from the mainly analytical nature of the rest of the book. It is intended simply to outline the types of pump or motor which may be used in hydraulic systems and to stress the similarities between the different types rather than their differences. Some analytical work is included dealing with flow pulsations mainly to indicate a feature which could be a source of vibration or oscillation in a completed system.

## 6.1 Types of Pump

Pumps employed to produce high pressures are almost always of the positive-displacement type. Pumps of the type in which kinetic energy is converted to pressure energy are not capable of producing large pressure increases efficiently because of the excessive fluid friction losses incurred at the very high velocities which would be needed within the pump (for example, an increase in pressure of 30 bar requires theoretically—neglecting losses—a velocity of about 90 m/s, see chapter 1).

Positive-displacement pumps which use pistons are suitable for the whole range of pressures likely to be encountered in oil hydraulic systems. A pump with pistons disposed in the axial direction is illustrated in figure 6.1;

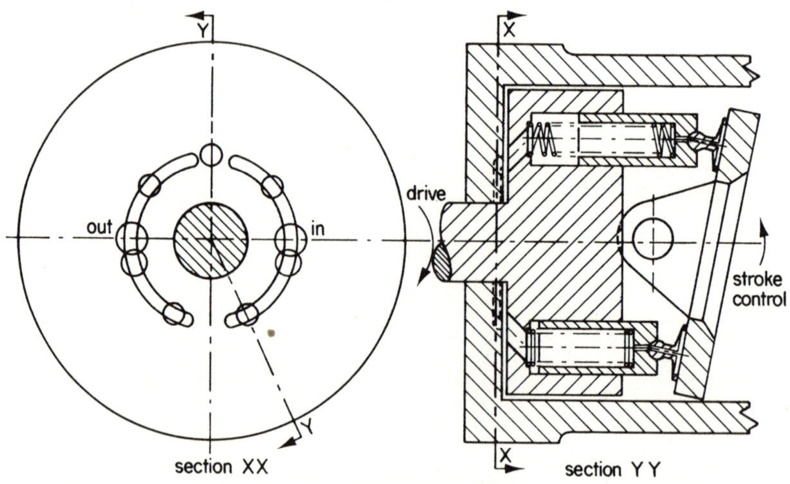


Figure 6.1 Swash plate unit

it has a 'swash plate' which may be tilted in order to change the flowrate. The plate may be tilted from the zero position to give flow in either direction and the angular position of the plate determines both the magnitude and

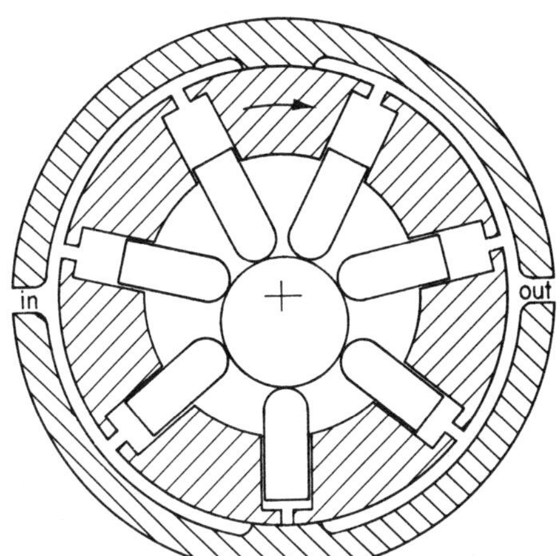


Figure 6.2  Radial piston

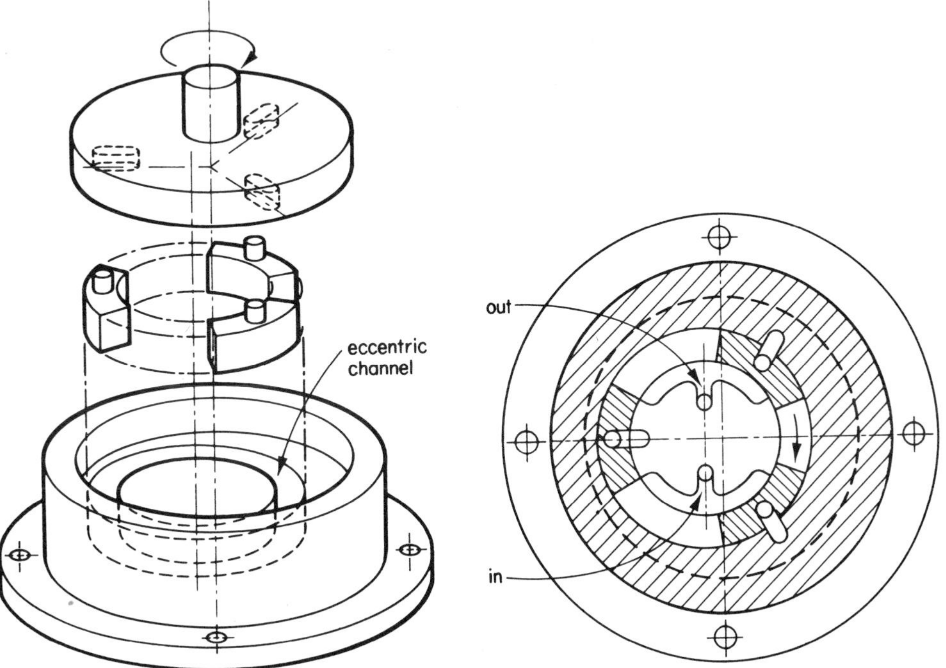


Figure 6.3  Moving segment

the direction of flow. This type of pump may also be used as a motor so long as positive-acting valves are not incorporated.

Another type of piston pump which has pistons disposed in the radial direction, and which may also be used as a motor, is illustrated in figure 6.2. It is of interest that the motions of each piston are the same as the motions of a piston actuated by a crank–connecting rod mechanism if the cam is circular although it is common practice to use a multi-lobe cam in this type of unit.

Another pump which is similar to a piston pump has segments which reciprocate along a circular channel of rectangular section. The delivery of this type of pump can be varied by altering the eccentricity of the channel (Savery pumps), see figure 6.3.

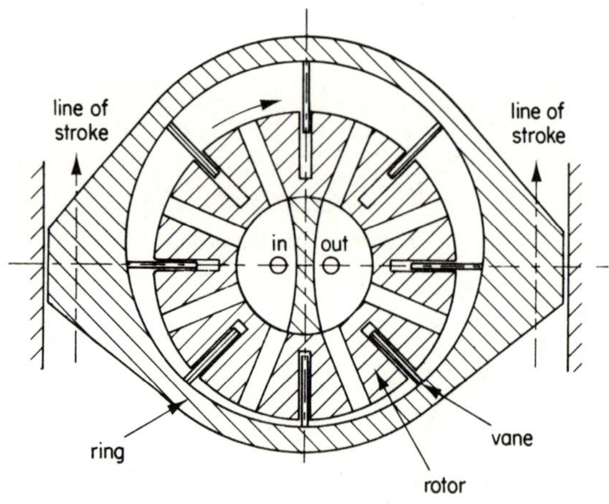


Figure 6.4  Variable vane

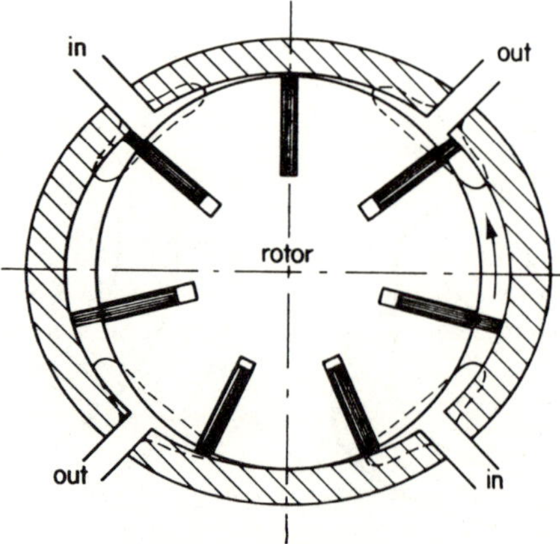


Figure 6.5  Balanced vane

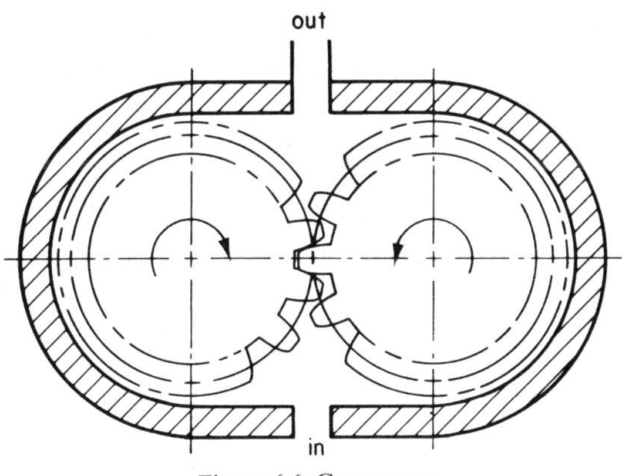



Figure 6.6  Gear pump

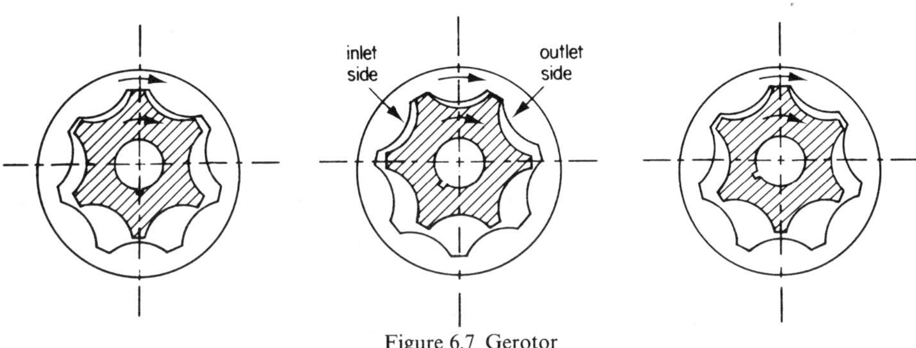



Figure 6.7  Gerotor

Positive-displacement pumps which employ vanes are normally limited to pressures below 170 bar. Both variable-delivery and fixed-delivery types are made. Variable-delivery pumps have a stator or control ring whose eccentricity can be altered as shown in figure 6.4. Vane pumps of the fixed-delivery type often have a double lobe as shown in figure 6.5 because this feature reduces the unbalanced radial forces acting on the rotor.

Gear pumps also operate by positive displacement of fluid from the inlet low-pressure side to the outlet at high pressure. Internal leakage normally limits their application to pressures of less than 140 bar. The pump with two gears illustrated in figure 6.6 transports pockets of fluid between adjacent teeth and return flow is prevented by the contact between teeth. Other types are similar to epicyclic gears and one example is illustrated in figure 6.7.

The design of different types of pump is considered in some detail by Khaimovich (1965) and by Korn (1969), see also Schlösser (1969).

## 6.2  *Flow Irregularities*

The flow from positive-displacement pumps is usually pulsating and the maximum variation in flowrate may be expressed as a percentage of the average flowrate. The number of piston (or vanes or other elements) employed is the main factor affecting the percentage irregularity as may be illustrated by considering the case of a piston pump in which the pistons are each actuated by a connecting rod coupled to a single crank. The rate of fluid flow from an individual cylinder is directly proportional to the instantaneous velocity of the piston in that cylinder, if fluid compressibility and leakage are neglected. With a crank rotating at constant speed $\Omega$, the velocity of any one piston at any instant will depend on the angular position of the crank at that instant. For the one piston shown in figure 6.8, the piston velocity will equal

$$\Omega r\left(\sin\theta - \frac{r\sin 2\theta}{2\ell}\right)$$

The flowrate contributed by each piston will be proportional to the velocity

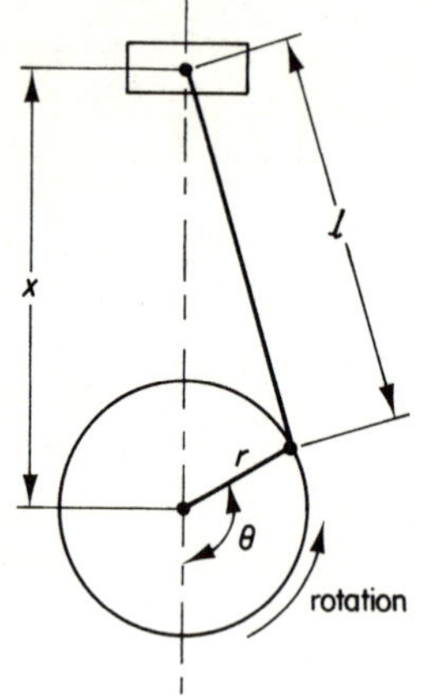

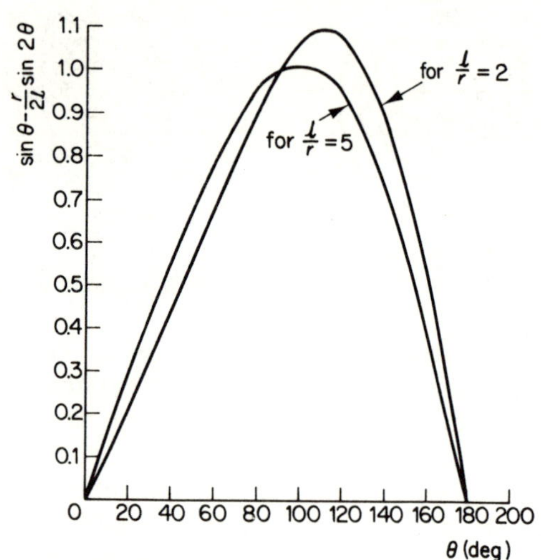


Figure 6.8

$$x = -r\cos\theta + \ell\left(1 - \frac{r^2}{\ell^2}\sin^2\theta\right)^{1/2}$$

$$= -r\cos\theta + \ell\left(1 - \frac{r^2}{2\ell^2}\sin^2\theta\right)$$

$$\approx \ell - r\cos\theta - \frac{r^2}{2\ell}\sin^2\theta$$

and where

$$\theta = \Omega t$$

$$\frac{\mathrm{d}x}{\mathrm{d}t} = \Omega r\sin\theta - \frac{r^2}{2\ell}\Omega\sin 2\theta$$

$$\frac{1}{\Omega r}\frac{\mathrm{d}x}{\mathrm{d}t} = \sin\theta - \frac{r}{2\ell}\sin 2\theta$$

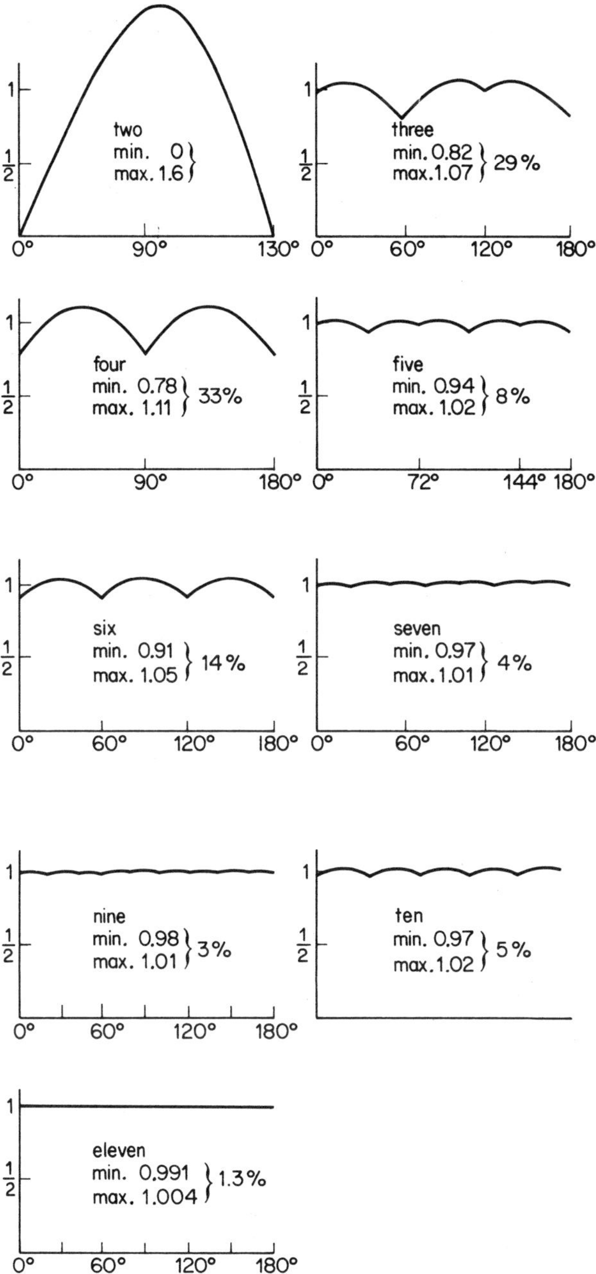



Figure 6.9 Flow irregularities for piston pumps with $\ell/r = 5$ (number of pistons stated)

of the piston at that instant and the total flowrate for a pump with $n$ pistons evenly disposed at intervals of $n/360$ degrees will be proportional to

$$\sum_{a_i=0}^{a_i=n-1} \sin\left(\theta + a_i \frac{2\pi}{n}\right) - \frac{r}{2\ell}\sin\left\{2\left(\theta + \frac{a_i 2\pi}{n}\right)\right\}$$

for positive values only—a negative value for any piston is associated with oil being drawn into the pump.

Numerical evaluations based on the above expression give the results plotted in figure 6.9 and summarised in figure 6.10.

The results illustrate the desirability of using an odd number of pistons.

It should be noted that flow pulsations and pressure pulsations are related according to the dynamic characteristics of the systems in which they occur. For example, a flow pulse in a short rigid pipeline would give a large pulse of pressure whereas the effect of the same flow pulse into a gas-loaded accumulator could well be undetectable.

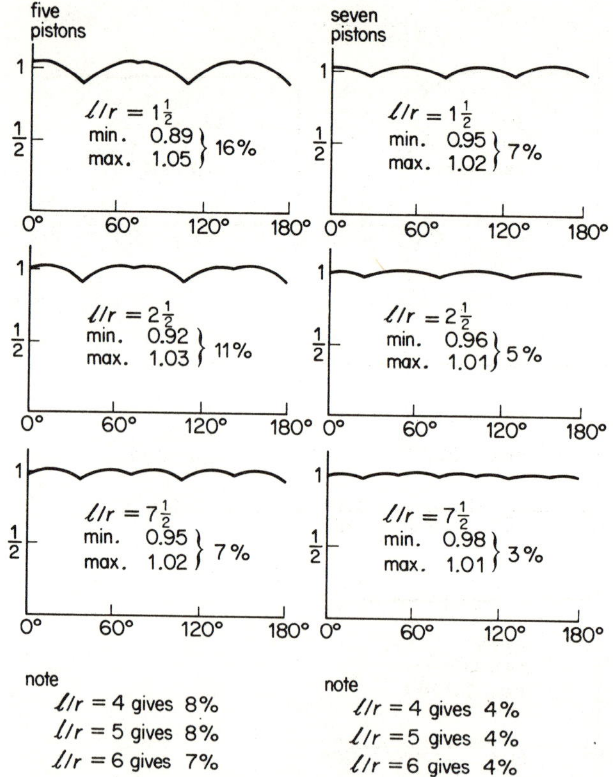


Figure 6.10 Flow irregularities for piston pumps with five or seven pistons
($\ell/r$ ratio stated)

### 6.3  *Constant-pressure Sources*

Many hydraulic systems need a supply of fluid whose pressure remains constant despite rapid fluctuations in the rate of flow demanded by the system. There are three main methods of satisfying this need.

The first method is to use a fixed-displacement pump of sufficient capacity to provide the maximum flowrate demanded by the system together with a pressure relief valve which bypasses excess fluid by throttling it from the supply pressure to the exhaust (or atmospheric) pressure. The method is cheap unless the relief valve regularly bypasses a large proportion of the flowrate when excessive heating will occur and involve the extra cost of some method of oil cooling.

A second method is to use a fixed-displacement pump which will not necessarily supply the maximum flowrate demanded by the system and to use it in conjunction with an accumulator. With this method, a non-return valve and an off-loading valve have to be included (possibly in one unit) in order to construct the type of circuit illustrated in figure 6.11.

The mode of operation is that, when the system is not demanding flow, then the accumulator accepts and stores fluid until its pressure rises sufficiently to actuate the off-loading valve. With the off-loading valve then open, the whole of the pump flowrate is recirculated to the oil reservoir (at very low pressure). Meantime the accumulator feeds fluid to the system and, as the accumulator empties, its pressure falls until it is sufficiently

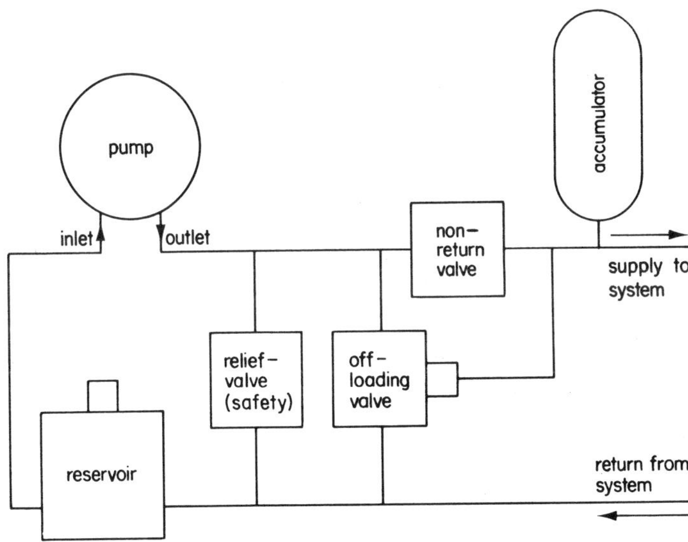


Figure 6.11

low to cause the off-loading valve to shut when the pumped supply is again fed to the accumulator and the system. This technique is probably best for a system which requires low flowrates for long periods interspersed with short periods of requiring high flowrates.

The third way of supplying variable flowrate of fluid at substantially constant pressure is to use a self-regulating variable-delivery pump. With this type of pump, a slight decrease of pressure causes an increase in flowrate and a slight increase in pressure causes a decrease in flowrate. The desired pressure can usually be preset and very small changes in pressure induce large changes in flowrate, although a possible disadvantage is the risk of the pump responding more slowly than other controls in the system. This method is popular for large power systems because the power absorbed by the pump varies continuously with the power demanded by the system and the losses are small.

# 7 Flow Through Valves

Most hydraulic servomechanisms or other high-performance systems rely for their operation on the metering of fluid through a valve. This chapter deals with a linearised method of analysis for 'four-way' valves. They are called this because they have four connections, one for the supply pressure, another for the exhaust, plus two control ports through both of which fluid may be metered, either from the supply to the system or from the system to the exhaust. (A valve may have five ports but these include either two pressure ports or two exhaust ports which can be connected together.) Such valves may be of the spool type as sketched in figure 7.1 or of the nozzle–flapper types. Other sorts of valves are used and the theory in this chapter is extended in appendix B to cover 'three-way' valves whilst a little information about poppet valves is given in appendix C.

Metering valves are never 'fully open' and their use is for accurately metering the flow of fluid through them. In the case of spool valves, longitudinal displacements of the spool are always small compared with the spool's diameter—for example, movements may be restricted to, say, 0.5 mm (0.02 in) in the case of a spool with diameter 5 mm (0.2 in). Similarly with a nozzle–flapper valve, the flapper only moves very small distances.

## 7.1 Four-way Spool Valves

A spool valve used for metering purposes controls flowrate by throttling. Each port in a valve which is partly closed off by a land on the spool becomes a control throttle.

The rate of flow of fluid through such a valve depends on the spool displacement from the null position $x$ (the valve opening) and on the pressures upstream and downstream of the valve. A convenient and representative pressure value to use when dealing with hydraulic systems is the pressure difference across the ram or motor which is being supplied by the valve. This will be termed $P_m$. One way of expressing the flowrate $q$ through a valve is

$$q = K_q x - K_c P_m \tag{7.1}$$

where $q$ is the volume flowrate of the oil (or other liquid); $K_q$ is the 'flow gain' (relating flow to valve opening if the pressures remain steady); $K_c$ is the 'pressure–flow coefficient' (relating flow to pressure if the valve opening remains steady); $x$ is the valve opening; $P_m$ is the pressure difference across the load.

Equation 7.1 implies that the flowrate is directly proportional to the valve opening and also directly proportional to the pressure drop across the valve. The equation, however, is an approximation and the approximate nature of equation 7.1 must be borne in mind when applying it.

### 7.1.1   *Critical Centre Valves*

Consider a four-way valve of the type sketched in figure 7.1 where the lands of the spool are exactly the same width as the annular ports of the valve body. In the central or null position the lands exactly cover the ports.

When the spool of such a valve is displaced (to the left, for example) by a distance $x$, then two annular orifices are formed of diameter $d_1$ and width $x$ (area $\pi d_1 x$) as indicated in figure 7.2.

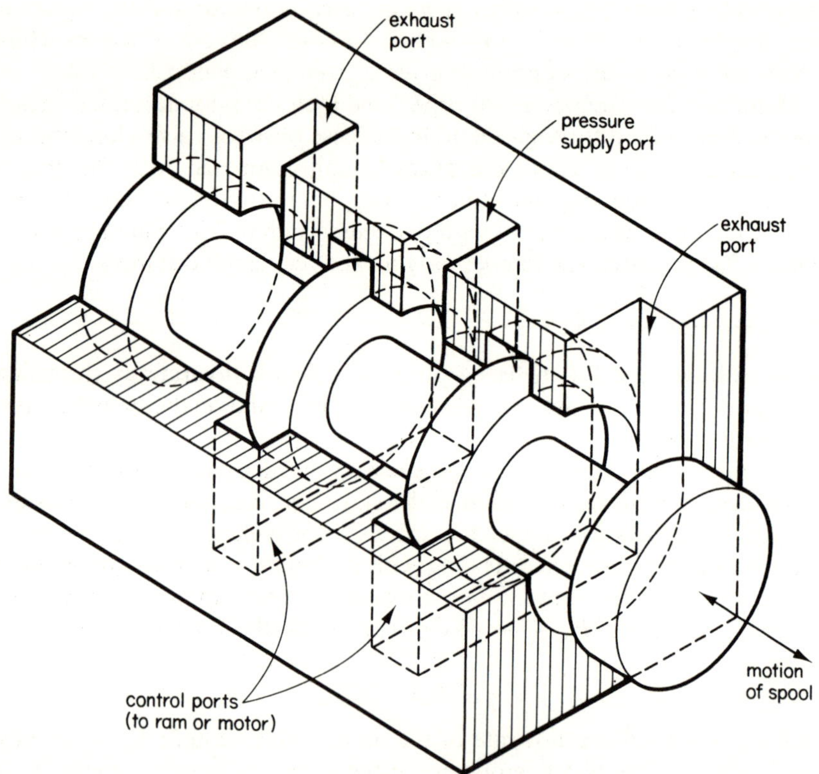


Figure 7.1 Sketch of spool valve: typical bore diameter 6 mm ($\frac{1}{4}$ in); typical radial clearance 12.7 $\mu$m (1/2000 in); typical axial displacement 0.6 mm (1/40 in)

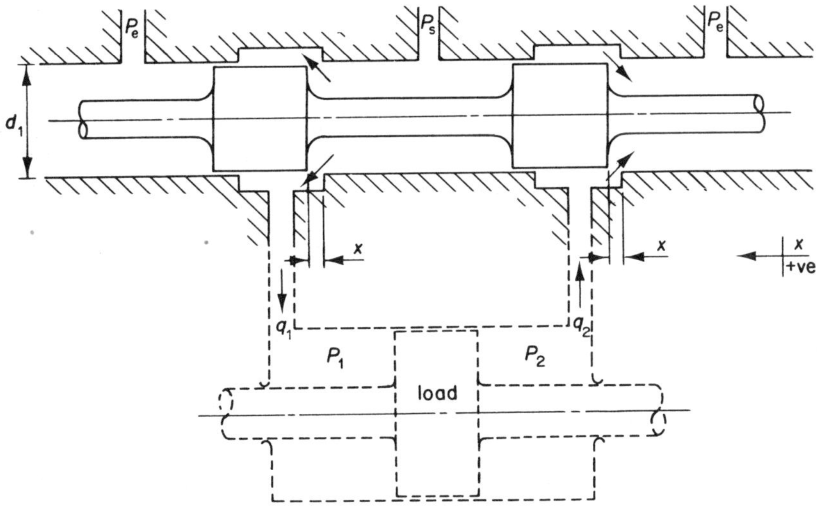


Figure 7.2 Four-way valve (diagrammatic)

### 7.1.2 *Flowrate Prediction*

Flow through sharp-edged orifices is predicted by applying the 'square root' law (see equations 1.1b and 1.2). Treating the annular orifices formed within the valve in this way gives

$$q_1 = C_d \pi d_1 x (P_s - P_1)^{1/2} \left(\frac{2}{\rho}\right)^{1/2}$$

and

$$q_2 = C_d \pi d_1 x (P_2 - P_e)^{1/2} \left(\frac{2}{\rho}\right)^{1/2}$$

which may be simplified by making certain assumptions or approximations to the actual situation with real valves. One assumption is that $q_1 = q_2$, implying that $P_s - P_1 = P_2 - P_e$. The other assumptions are that the supply pressure $P_s$ is constant and that the exhaust pressure is negligible (or $P_e = 0$). By introducing the term $P_m = P_1 - P_2$, we have $P_s - P_m = 2P_2$ and $P_s + P_m = 2P_1$. The two equations can then be written

$$q = C_d \pi d_1 x (P_s - P_m)^{1/2} \left(\frac{1}{\rho}\right)^{1/2} \quad (= q_1 = q_2) \tag{7.2}$$

With the type of configuration illustrated in figures 7.1 and 7.2, it is usually accepted that $C_d \approx \frac{5}{8}$ and is constant. Also for oil $\rho \approx 870$ kg/m$^3$ (about 1/32 lb/in$^3$), whence equation 7.2 can be rewritten

$$q \approx 6.7\pi d_1 x (P_s - P_m)^{1/2} \tag{7.3}$$

(equation 7.3 relates to SI units with pressures in *bars*, that is $q$ in m$^3$/s, $d_1$ and $x$ in m). (For in lb s units $q \approx 70\pi d_1 x (P_s - P_m)^{1/2}$.)

The assumption that the valve openings can be treated as orifices presupposes that $x$ is small compared with $d_1$ so that the pressure drop across each orifice will be significant compared with $P_s$. In practice the pressure difference across the load $P_m$ rarely exceeds $\frac{2}{3}P_s$, allowing further simplification of equation 7.2 or 7.3. If $P_m < \frac{2}{3}P_s$, then less than 10% error is involved in using the (binomial) approximation

$$(P_s - P_m)^{1/2} = (P_s)^{1/2}\left(1 - \frac{P_m}{P_s}\right)^{1/2} \approx (P_s)^{1/2}\left(1 - \frac{1}{2}\frac{P_m}{P_s}\right) \tag{7.4}$$

so that equation 7.3 becomes

$$q = 6.7\pi d_1 x (P_s)^{1/2} - 6.7\pi d_1 x (P_s)^{1/2}\frac{1}{2P_s}P_m \tag{7.5}$$

And equation 7.5 is similar in form to equation 7.1 with

$$K_q = 6.7\pi d_1 (P_s)^{1/2} \tag{7.6}$$

$$K_c = \frac{6.7\pi d_1 x (P_s)^{1/2}}{2P_s} \tag{7.7}$$

(equations 7.6 and 7.7 refer to four-way critical centre spool valves and apply to SI units with pressures in *bars*). (For British units 70 replaces 6.7 in both equations.)

The above analysis predicts that the 'flow gain' $K_q$ can be treated as a constant for a particular valve and supply pressure but that the 'pressure–flow coefficient' $K_c$ will vary with the valve opening $x$. The variations of $K_c$ are of minor significance for linear analysis purposes as will be mentioned later.

The flow gain $K_q$ of a valve may be predicted from equation 7.6. Or, alternatively, it may be estimated from the known 'rating' of a particular (metering or servo) valve if the maximum spool displacement for the valve is also known. In the latter case, the 'rating' will probably be associated with a given load pressure, most probably two-thirds of the supply pressure when a notional flowrate of $(3)^{1/2}$ times the rating would be invoked to evaluate $K_q$. Let us consider the following numerical example. A valve for use at a supply pressure of 210 bar has a rating of 0.53 l/s (7 gal/min or 32.2 in$^3$/s) when the valve pressure drop is one-third of the supply pressure or 70 bar, that is 35 bar (507 lb/in$^2$) at each port or $P_s = 210$ bar (3046 lb/in$^2$), $P_1 = 175$ bar (2538 lb/in$^2$), $P_2 = 35$ bar (507 lb/in$^2$) and $P_e = 0$ when $P_m$ is 140 bar (2031 lb/in$^2$). The maximum spool travel is 0.6 mm (0.024 in) and the spool diameter 5 mm (0.197 in). Estimate $K_q$.

(a) $K_q = 6.7\pi d_1 (P_s)^{1/2} = 6.7 \times \pi \times 5 \times 10^{-3}(210)^{1/2} = 1.53$

(b) $K_q = \dfrac{0.53 \times 10^{-3} \times (3)^{1/2}}{0.6 \times 10^{-3}} = 1.53$

(for (b), $0.53 \times 10^3 \times (3)^{1/2} \mathrm{m}^3/\mathrm{s}$ is the notional flowrate for 105 bar pressure drop at each port when $P_m = 0$).

In British units

(a) $K_q = 70 \times \pi \times 0.197 \times (3046)^{1/2} \approx 2400$

(b) $K_q = \dfrac{32 \times 2 \times (3)^{1/2}}{0.024} \approx 2300$

The flow gain $K_q$ will not be exactly constant for a real valve particularly (but not only) because the spool lands will never exactly match the annular ports in the valve body. Actual test results with a constant pressure drop across the valve ports would show variations, particularly near the central or null position of the spool, such as those illustrated in figure 7.3. The flow gain $K_q$ is the slope of the appropriate line in figure 7.3, which can double its value near null with negative lap. The magnitude of $K_q$ is the most important parameter of a valve and often also of any system incorporating the valve.

The 'pressure–flow coefficient' $K_c$ is not a constant for a particular valve

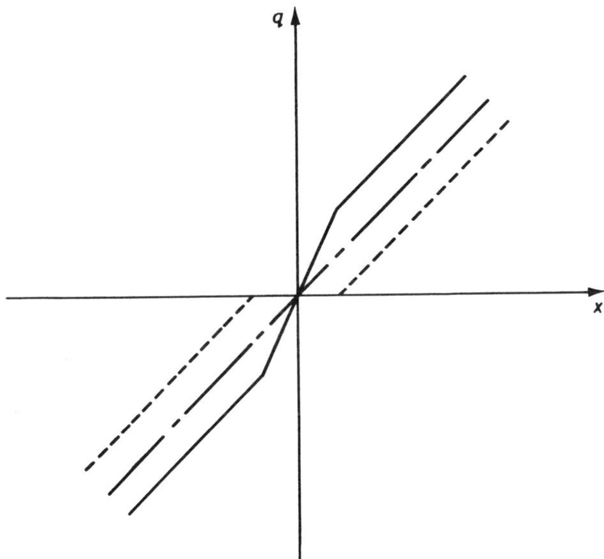


Figure 7.3 Flowrates versus valve displacement (for constant pressure drop): —·—ideal (zero lap); ————underlap (open centre); ————overlap

and supply pressure but increases with valve opening. Its smallest value, usually the one most relevant in system design, occurs near null ($x \approx 0$) where the calculated value for a zero-lap valve would be zero. In practice, a value of zero would not occur mainly because of the radial clearance between the spool and valve body which gives a flow path for the fluid at the null position. A practical method of finding the null value (termed $K_{c0}$) is to make tests on the particular valve at spool displacements near zero with the two control ports blocked by pressure gauges. The results will show how the two pressures vary with valve displacement near the null position. A result of the order of 3000 bar/mm may be obtained (see Merritt (1967), in which $10^6$ (lb/in$^2$)/in is quoted)—call this $K_{p0}$. With the flow gain $K_q$ known, the result is obtained as $K_{c0} = K_q/K_{p0}$ in units of (m$^3$/s)/(N/m$^2$) or (in$^3$/s)/(lb/in$^2$). A valve with large spool–bore clearance or rounded land edges or with negative lap gives smaller values of $K_{p0}$, hence larger values of $K_{c0}$. A system incorporating a valve depends on the value of $K_c$ for the damping of oscillations. $K_c$ increases with $x$ and so system damping is greater at any instant when the valve is open than at other instants when the valve is at or near its closed position.

### 7.1.3   *Open Centre Type (Underlapped Four-way Valve)*

A valve in which the lands of the spool never completely cover the ports of the valve body is said to be underlapped (or to have negative lap). The

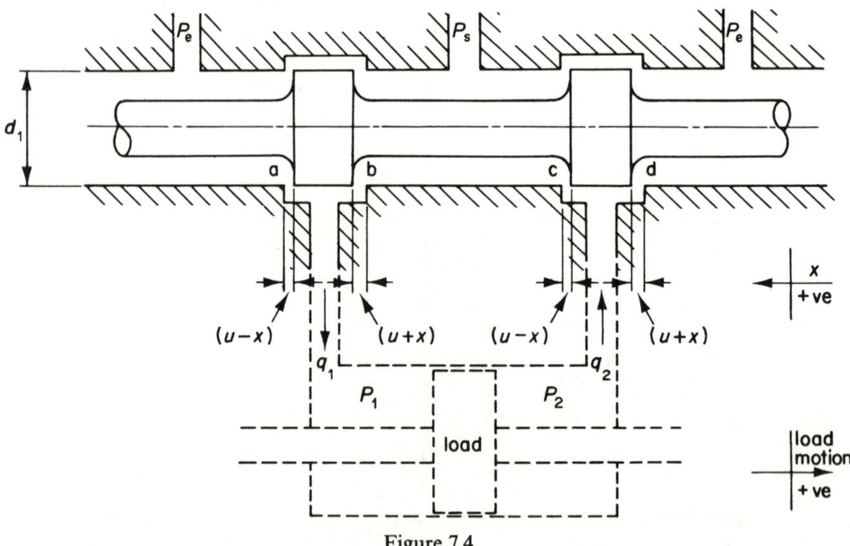


Figure 7.4

same equation is again employed, namely

$$q = K_q x - K_c P_m \tag{7.1}$$

and values of $K_q$ and $K_c$ can be obtained for operation *in the underlap region only* (that is no port edge ever being completely shut off by the spool).

Referring to figure 7.4, a displacement of $x$ (say to the left) unbalances the (assumed) symmetry of the ports. Two of the annular orifices increase in width from $u$ to $u + x$ and two decrease from $u$ to $u - x$.

The flowrates $q_1$ and $q_2$ may be estimated as follows:

$$q_1 = C_d \pi d_1 (u + x)(P_s - P_1)^{1/2} \left(\frac{2}{\rho}\right)^{1/2} - C_d \pi d_1 (u - x)(P_1 - P_e)^{1/2} \left(\frac{2}{\rho}\right)^{1/2}$$

and

$$q_2 = C_d \pi d_1 (u + x)(P_2 - P_e)^{1/2} \left(\frac{2}{\rho}\right)^{1/2} - C_d \pi d_1 (u - x)(P_s - P_2)^{1/2} \left(\frac{2}{\rho}\right)^{1/2}$$

and assuming $q_1 = q_2$ and $P_s$ remains constant and $P_e = 0$, so that $P_s = P_1 + P_2$ and writing $P_m = P_1 - P_2$, we obtain

$$q = C_d \pi d_1 \left\{ (u + x)(P_s - P_m)^{1/2} - (u - x)(P_s + P_m)^{1/2} \right\} \left(\frac{1}{\rho}\right)^{1/2} \tag{7.8}$$

which may be approximated as

$$q = C_d \pi d_1 \left\{ (u + x)(P_s)^{1/2}\left(1 - \frac{1}{2}\frac{P_m}{P_s}\right) - (u - x)(P_s)^{1/2}\left(1 + \frac{1}{2}\frac{P_m}{P_s}\right) \right\} \left(\frac{1}{\rho}\right)^{1/2}$$

$$= C_d \pi d_1 \left(\frac{1}{\rho}\right)^{1/2} (P_s)^{1/2} 2x - C_d \pi d_1 u \left(\frac{1}{\rho}\right)^{1/2} \frac{(P_s)^{1/2}}{P_s} P_m \tag{7.9}$$

The numerical value of $C_d (1/\rho)^{1/2}$ is about 6.7 for SI units with pressures in bars (and 70 for in lb s units) and equation 7.9 has the same form as equation 7.1 with

$$K_q = 13.4 \pi d_1 (P_s)^{1/2} \tag{7.10}$$

and

$$K_c = \frac{6.7 \pi d_1 u (P_s)^{1/2}}{P_s} \tag{7.11}$$

for four-way open centre SI units. Note also that $K_p$ (that is $K_q/K_c$) equals $2P_s/u$, where $u$ is the distance between each of the four land edges and port edges (that is the four identical port openings) when the spool is centred.

The values refer to operation within the underlap region. Outside this region these valves act as critical centre valves with only two active ports.

Note particularly that the flow gain $K_q$ is double that for a comparable critical centre valve (figure 7.3) in the underlap region. Note also the significant leakage flow when the valve is centred (leakage flow at null when the load flow $q$ is zero becomes $13.4\pi d_1 u(P_s)^{1/2}$).

Thorough analyses of lapped valve and associated ram systems have been given by Shearer (1954) and Royle (1961).


## 7.2   Three-way Spool Valves

Three-way valves have only one critical length dimension which helps to ease manufacture. However, they cannot be used for motors requiring flow reversal and are usually applied to differential rams as indicated in appendix B.


## 7.3   Flapper-nozzle Valves

Commonly used as the first stage of two-stage servo valves. Nozzles and the fixed upstream orifices used with them are made with diameters ($d_n$ and $d_o$) in the range 0.2 mm to 0.8 mm and the distance $x_0$ between each nozzle and the flapper in a double valve is often less than 0.2 mm. Each nozzle and orifice is as nearly a sharp-edged orifice as possible and treated as such for analytical purposes. The curtain area formed by the flapper at a nozzle exit modulates the control pressure caused by the fixed upstream orifice.

Consider the double-nozzle–flapper valve indicated in figure 7.5.

We assume that the valve has a balance condition such that $x = 0$ and $P_m = 0$ when $q = 0$. This occurs with the pressure downstream of each of the fixed orifices equal to $P_s/2$ and when the flow through each orifice equals that through each nozzle, that is

$$q_1 \text{ (steady state)} = q_3 \text{ (steady state)} = C_{do}A_o\left(\frac{2}{\rho}\right)^{1/2}\left(P_s - \frac{P_s}{2}\right)^{1/2}$$

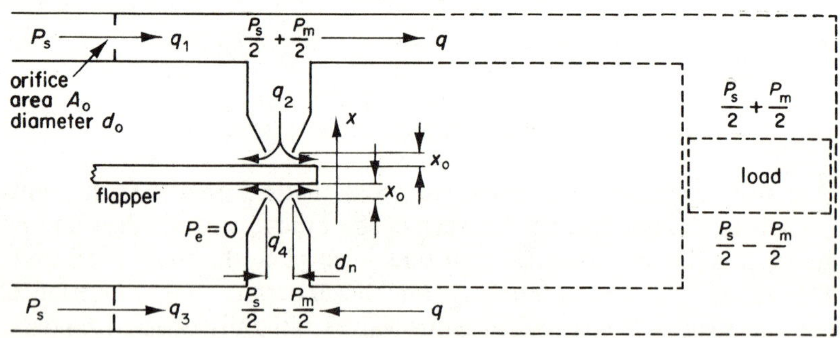


Figure 7.5 Double-nozzle–flapper valve

and

$$q_2 \text{ (steady state)} = q_4 \text{ (steady state)} = C_{dn}\pi x_0 \left(\frac{2}{\rho}\right)^{1/2}\left(\frac{P_s}{2}\right)^{1/2}$$

and $q_1 = q_3 = q_2 = q_4$ which also implies that the orifice size and the curtain area in the null position are approximately equal or

$$C_{do}A_o = C_{dn}\pi d_n x_0$$

(and it is suggested by Morse (1963) that $d_n = 1\frac{1}{2}d_o$ and $d_n = 8x_0$ are feasible design ratios; although these give $\pi d_n x_0 / A_o$ as about 1.1 instead of unity, $C_{do}$ may be different from $C_{dn}$).

Considering the valve not in balance, that is $x$ has some value as does $P_m$, we have (using $C_{do}A_o (2/\rho)^{1/2} = C_{dn}x d_n x_0 (2/\rho)^{1/2} = K_n$)

$$q = q_1 - q_2 = K_n\left\{P_s - \left(\frac{P_s}{2} + \frac{P_m}{2}\right)\right\}^{1/2} - K_n\left(\frac{x_0 - x}{x_0}\right)\left(\frac{P_s}{2} + \frac{P_m}{2}\right)^{1/2}$$

$$q = -q_3 + q_4 = -K_n\left\{P_s - \left(\frac{P_s}{2} - \frac{P_m}{2}\right)\right\}^{1/2} + K_n\left(\frac{x_0 + x}{x_0}\right)\left(\frac{P_s}{2} - \frac{P_m}{2}\right)^{1/2}$$

and using the binomial approximations

$$\left(\frac{P_s}{2} \pm \frac{P_m}{2}\right)^{1/2} = \left(\frac{P_s}{2}\right)^{1/2}\left(1 \pm \frac{1}{2}\frac{P_m}{P_s}\right)$$

we find

$$q = K_n\left(\frac{P_s}{2}\right)^{1/2}\left(1 - \frac{1}{2}\frac{P_m}{P_s}\right) - K_n\left(\frac{P_s}{2}\right)^{1/2}\left(1 + \frac{1}{2}\frac{P_m}{P_s}\right)$$

$$+ K_n\frac{x}{x_0}\left(\frac{P_s}{2}\right)^{1/2}\left(1 + \frac{1}{2}\frac{P_m}{P_s}\right)$$

$$q = -K_n\left(\frac{P_s}{2}\right)^{1/2}\left(1 + \frac{1}{2}\frac{P_m}{P_s}\right) + K_n\left(\frac{P_s}{2}\right)^{1/2}\left(1 - \frac{1}{2}\frac{P_m}{P_s}\right)$$

$$+ K_n\frac{x}{x_0}\left(\frac{P_s}{2}\right)^{1/2}\left(1 - \frac{1}{2}\frac{P_m}{P_s}\right)$$

Adding and dividing by 2

$$q = -K_n\left(\frac{P_s}{2}\right)^{1/2}\frac{P_m}{P_s} + K_n\frac{x}{x_0}\left(\frac{P_s}{2}\right)^{1/2}$$

which is in the form of

$$q = K_q x - K_c P_m \tag{7.1}$$

where

$$K_q = \frac{K_n}{x_0}\left(\frac{P_s}{2}\right)^{1/2} = C_{dn}\pi d_n\left(\frac{2}{\rho}\right)^{1/2}\left(\frac{P_s}{2}\right)^{1/2} \tag{7.12}$$

and

$$K_c = \frac{K_n}{P_s}\left(\frac{P_s}{2}\right)^{1/2} = \frac{C_{dn}\pi d_n x_0}{P_s}\left(\frac{2}{\rho}\right)^{1/2}\left(\frac{P_s}{2}\right)^{1/2} \tag{7.13}$$

$$= \frac{C_{do}A_o}{P_s}\left(\frac{2}{\rho}\right)^{1/2}\left(\frac{P_s}{2}\right)^{1/2}$$

for a four-way flapper.

Three-way flapper valves are also used; they give slightly less linear relations; their null balance is affected by supply pressure changes; but they are simpler to make than four-way valves.


## Problems

1 A four-way valve with full periphery annular ports has a 6 mm (0.236 in) diameter spool and it may be assumed that the spool lands fully cover the valve ports in the zero or midposition. Estimate the flowrate through one port when the pressure drop across it is 70 bar (1015 lb/in$^2$) for every millimetre (or inch) of spool displacement.

(1.49 (1/s)/mm, 2315 (in$^3$/s)/in.)


2 What would be the flow coefficient $K_q$ of the above valve if it were used as part of a servo system having oil supply pressure (a) 140 bar, (b) 210 bar?

((a) 1.49, (b) 1.83 (m$^3$/s)/mm; (a) 2315, (b) 2843 (in$^3$/s)/in.)


3 A three-way spool valve with half the annular periphery of the valve port blocked off and spool diameter 9 mm is used in a system supplied with oil at 120 bar pressure. The 'half-area' piston has areas of 0.004 m$^2$ and 0.002 m$^2$ and a maximum required velocity 0.3 m/s. Estimate the maximum spool displacement required (about 2 mm, allowing for one-third pressure drop through valve).


4 In a 240 bar (3480 lb/in$^2$) servo system employing a four-way valve, valve underlap is used to assist in damping system oscillations. The valve has a 4 mm (0.157 in) diameter spool, full periphery ports and nominal underlap of 0.0127 mm (0.0005 in). Estimate the pressure–flow coefficient $K_c$ for the valve.

(6.9 × 10$^{-8}$ (m$^3$/s)/bar; 2.9 × 10$^{-4}$ (in$^3$/s)/(lb/in$^2$).)

# 8 Valve-controlled Systems

This chapter deals with the linear method of analysing hydraulic devices which are controlled by using metering valves. Special attention is given to pistons controlled by metering valves of the four-way type but the method is equally applicable to systems employing motors or other types of valve (for three-way valves, see appendix B). The aim of these linear or small perturbation analyses is to derive mathematical models which give reasonably accurate representations of practical system behaviour without being unduly complicated.

The analyses in this chapter make use of the relations developed in chapter 7 and of the procedures described in chapters 2 and 3. Governing equations are derived by combining the simultaneous equations concerned respectively with a balance of flowrates and a balance of forces. The final sections of this chapter concern servomechanisms and draw on the methods of chapter 5, particularly the graphical procedure involving Bode diagrams, to demonstrate the selection of system parameters in order to meet given design specifications.

## 8.1 Four-way Valve System

A piston controlled by a four-way valve is illustrated in figure 8.1.

Referring to figure 8.1, a displacement of the valve spool to the right, for example, would cause fluid from the supply to be metered into the left-hand chamber. At the same time it would cause fluid to be metered from the right-hand chamber to exhaust. An increase in pressure in one chamber and a corresponding decrease in the other would give rise to a force tending

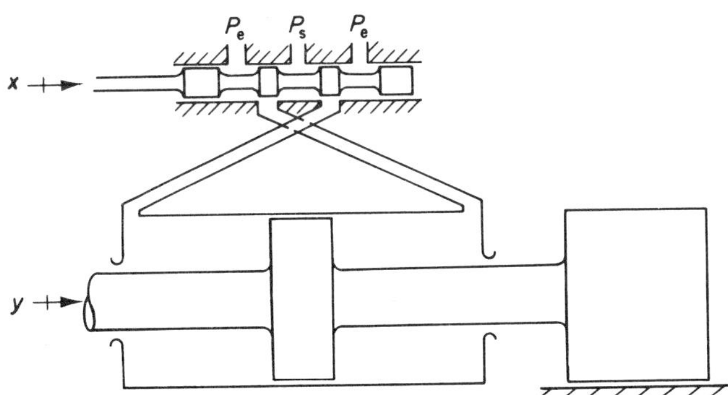


Figure 8.1 Cylinder with four-way valve

to displace the piston and its load. We shall now consider the load motions induced by moving the valve spool and assume that the supply pressure $P_s$ remains constant throughout.

## 8.2   Pure Inertia Analysis

To begin with, the load will be assumed to be simply a mass with no frictional forces involved, no externally applied forces and with no leakage occurring. It will also be assumed that the flow entering through the valve from the supply exactly equals that leaving through the valve to exhaust (volume flowrates being considered throughout). The piston will be assumed to be in its central position (where the oil 'spring' has its minimum stiffness). Small motions about the centred position are assumed for both the piston and the spool, that is this is a small perturbation analysis.

The volume flowrate to and from the cylinder may be written

$$q = K_q x - K_c P_m \tag{8.1}$$

(repeat of equation 7.1) where $P_m = P_1 - P_2$ for this type of system.

This may be equated to either the flowrate entering one port of the valve or leaving the other; this flowrate causes (a) motion and (b) compression of fluid

$$q = A \frac{\mathrm{d}y}{\mathrm{d}t} + \frac{V_t/2}{\beta} \frac{\mathrm{d}P_1}{\mathrm{d}t} = A \frac{\mathrm{d}y}{\mathrm{d}t} - \frac{V/2}{\beta} \frac{\mathrm{d}P_2}{\mathrm{d}t}$$

where $V_t$ is the *total* cylinder volume. Note that inherent in the 'equal flow' assumption is that $P_1$ increases at the same rate as $P_2$ decreases. The equations may be combined to give

$$q = A \frac{\mathrm{d}y}{\mathrm{d}t} + \frac{V_t}{4\beta} \frac{\mathrm{d}P_m}{\mathrm{d}t} \tag{8.2}$$

The net force on the piston produces acceleration

$$A(P_1 - P_2) = A P_m = M \frac{\mathrm{d}^2 y}{\mathrm{d}t^2} \tag{8.3}$$

and equation 8.3 may be substituted in the second term of 8.1 and, after differentiating, in the second term of 8.2, allowing 8.1 and 8.2 to be equated giving

$$\frac{K_q}{A} x = \frac{V_t}{4\beta} \frac{M}{A^2} \frac{\mathrm{d}^3 y}{\mathrm{d}t^3} + K_c \frac{M}{A^2} \frac{\mathrm{d}^2 y}{\mathrm{d}t^2} + \frac{\mathrm{d}y}{\mathrm{d}t} \tag{8.4}$$

(for the four-way valve system of figure 8.1 with pure inertia load) or as the operational relation between piston speed $\mathrm{D}y$ and valve displacement $x$

$$\frac{\mathrm{D}y}{x} = \frac{K_q/A}{(1/\omega_\mathrm{h}^2)\mathrm{D}^2 + (2\zeta/\omega_\mathrm{h})\mathrm{D} + 1} \qquad (8.5)$$

where

$$\omega_\mathrm{h}^2 = \frac{4\beta}{V_\mathrm{t}}\,\frac{A^2}{M}$$

(for a four-way valve system) and

$$\frac{2\zeta}{\omega_\mathrm{h}} = K_\mathrm{c}\,\frac{M}{A^2}$$

whence

$$\zeta = \frac{K_\mathrm{c}}{2A}\left(\frac{\beta M}{V_\mathrm{t}}\right)^{1/2}$$

note that $\mathrm{D} \equiv \mathrm{d}/\mathrm{d}t$. The damping factor $\zeta$ is directly dependent on the valve pressure–flow coefficient $K_\mathrm{c}$ in the absence of other damping.

### 8.2.1 *Analysis with Friction and Leakage*

Shunt leakage may be introduced across a ram or motor to improve damping although this has a deleterious affect on steady state accuracy, see chapter 10, section 10.4. Viscous friction may be present, or introduced also for damping. A four-way valve system will be considered with leakage flowrate directly proportional to pressure drop $P_1 - P_2$ across, for example, a small-bore tube. With such a tube, flowrates would actually be viscosity dependent (hence temperature dependent) to avoid which a sharp-edged orifice restriction may be preferred. Orifice effects are more difficult to predict analytically. In real systems and in analogue computer simulations, the orifice alternative may be considered but here we take leakage flow $q_\mathrm{L}$ as

$$q_\mathrm{L} = L(P_1 - P_2) = LP_\mathrm{m}$$

where $L$ is a constant.

Friction will be assumed as purely viscous. In real systems, other types of friction occur and, for analogue computer simulations, other frictional forces may be included. For this analysis the frictional force is taken as

$$f\,\frac{\mathrm{d}y}{\mathrm{d}t}$$

where $f$ is a constant.

The analysis is similar to that of the pure inertia case. Equation 8.1 (that is $q = K_q x - K_\mathrm{c}P_\mathrm{m}$) is retained but equations 8.2 and 8.3 are modified.

The flow entering one port or leaving the other can be considered to have three components: the first provides velocity of the piston, the second is the compressibility flow and the third is leakage, viz.

$$q = A\,\frac{dy}{dt} + \frac{V_t/2}{\beta}\,\frac{dP_1}{dt} + L(P_1 - P_2)$$

or for the other port

$$q = A\,\frac{dy}{dt} - \frac{V_t/2}{\beta}\,\frac{dP_2}{dt} + L(P_1 - P_2)$$

giving

$$q = A\,\frac{dy}{dt} + \frac{V_t}{4\beta}\,\frac{dP_m}{dt} + LP_m \tag{8.2a}$$

The net force on the piston not only causes acceleration but also has to overcome friction whence

$$AP_m = M\,\frac{d^2y}{dt^2} + f\,\frac{dy}{dt} \tag{8.3a}$$

and equation 8.3a may be substituted in the second term of 8.1 and the third term of 8.2a and, after differentiating in the second term of 8.2a, allowing 8.1 and 8.2a to be equated giving

$$\frac{K_q x}{A} = \frac{V_t}{4\beta}\,\frac{M}{A^2}\,\frac{d^3y}{dt^3} + \left(L + K_c + \frac{fV_t}{4\beta M}\right)\frac{Md^2y}{A^2 dt^2}$$

$$+ \left(1 + \frac{Lf}{A^2} + \frac{Kf}{A^2}\right)\frac{dy}{dt} \tag{8.4a}$$

(for a four-way valve system with inertia, viscous friction and laminar leakage).

Now for cases of $(L + K_c)f/A^2 \ll 1$, the operational relation between piston speed $Dy$ and valve displacement $x$ may be written

$$\frac{Dy}{x} \approx \frac{K_q/A}{(1/\omega_h{}^2)D^2 + (2\zeta'/\omega_h)D + 1} \tag{8.5a}$$

where

$$\omega_h{}^2 = \frac{4\beta}{V_t}\,\frac{A^2}{M}$$

as before (re equation 8.5) but

$$\frac{2\zeta'}{\omega_h} = \left(L + K_c + \frac{fV_t}{4\beta M}\right)\frac{M}{A^2}$$

showing that the shunt leakage and the viscous friction both help to increase the damping coefficient. (The ratio between $\zeta$ re equation 8.5 and $\zeta'$ in equation 8.5a is

$$\frac{\zeta'}{\zeta} = \frac{L + K_c + f V_t/4\beta M}{K_c}$$

and it can be shown alternatively that, with the same shunt leakage but with no viscous friction, $\zeta$ increases in the ratio $(L + K_c)/K_c$.)

## 8.3   Valve Position Servos

Follow-up systems, particularly for rectilinear motions, commonly employ servo valves with mechanical feedback linkages. Examples occur, for example, on copying lathes and steering boosters.

One such position controller sketched in figure 8.2 uses a four-way valve (for other systems, see figure 8.4 and appendix B). We shall consider the dynamic behaviour of such a moving body type of system.

A displacement of the input $\theta_i$ to the right in figure 8.2 would cause the valve to meter fluid from the supply into the right-hand chamber and from the left-hand chamber to exhaust, thereby causing the cylinder to move towards the right and to bring the valve to its closed position. As a first (steady state) approximation, a movement $\theta_i$ is matched by an equal movement $\theta_o$—the force applied to cause the movement being negligible compared with the force available to move the load. The response is not instantaneous and is likely to be oscillatory whence the need for dynamic analysis.

For analysis it will be assumed that the supply pressure $P_s$ remains

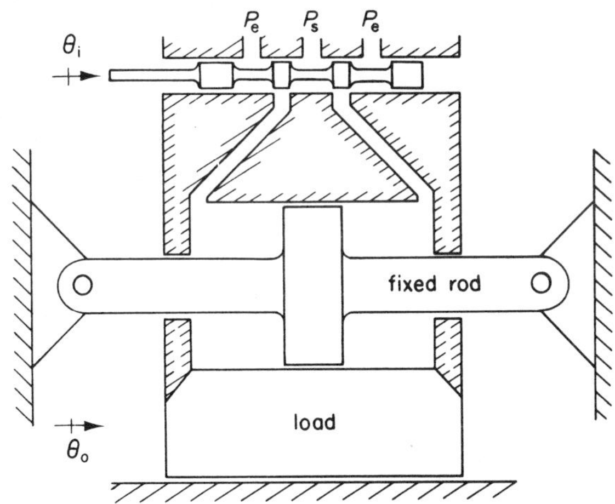


Figure 8.2 Position control system

constant at all times, owing probably to the presence of an accumulator in the supply line of sufficient capacity to ensure a negligibly small fluctuation at maximum transient flowrate.

### 8.3.1 *The 'Velocity Constant'*

One characteristic or parameter of the system is associated with moving the input horizontally at constant velocity. The valve must remain open a sufficient distance to provide the flowrate needed to keep the load moving at this velocity. There is thus a 'velocity error', the moving cylinder always lagging behind the input as regards position when both are moving at a steady speed, by an amount which depends both on the speed and also on the ratio between the valve's flow coefficient and the net (piston) area.

In the simplest analysis with no frictional forces or externally applied forces on the load, the flowrate would depend only on the valve opening according to $q = K_q x$ (with the second term of equation 8.1 equal to zero). For a steady speed of $\Omega$ (m/s for example), taking the flowrate equal to $A\Omega$ we have

$$K_q x_s = A\Omega \text{ or } x_s = \frac{\Omega}{K_q/A}$$

where $x_s$ is the steady 'velocity error' or the steady state difference between $\theta_i$ and $\theta_o$ necessary to sustain velocity $\Omega$. The term $K_q/A$ is called the 'velocity constant' of the system (possibly in units (m/s)/m or (in/s)/in).

### 8.3.2 *Governing Equation*

A differential equation is required to represent the system. A method of analysis similar to that used for the open loop system shown in figure 8.1 is relevant. In fact all the equations developed for the open loop system are directly applicable to this closed loop system and need not be repeated.

The flowrate through the valve is as shown in equation 8.1. The motion is of the cylinder rather than the piston but otherwise equation 8.2 applies. Assuming a pure inertia load, equation 8.3 applies direct (reading $\theta_o$ for $y$), noting that $M$ represents the total moving mass including in this case the cylinder and valve body. The individual equations apply direct so the combined equation 8.4 must also apply. In equation 8.4 the symbol $y$ should be replaced by $\theta_o$.

The basic difference between the system sketched in figure 8.2 and that in figure 8.1 is that the displacements of the load and valve spool are no longer independent. For the system of figure 8.2, these are related by the equation $x = \theta_i - \theta_o$.

By substituting $x = \theta_i - \theta_o$ in equation 8.4 we obtain the governing equation for the system shown in figure 8.2 which is

$$\frac{K_q}{A}\,\theta_i = \frac{V_t}{4\beta}\frac{M}{A^2}\frac{d^3\theta_o}{dt^3} + K_c\,\frac{M}{A^2}\frac{d^2\theta_o}{dt^2} + \frac{d\theta_o}{dt} + \frac{K_q}{A}\,\theta_o \tag{8.6}$$

(for the four-way valve system of figure 8.4 but without shunt leakage and with pure inertia load).

Where the load comprises inertia and viscous friction and in the presence of shunt leakage (of rating $L$—probably in units $(m^3/s)/(N/m^2)$ or $(in^3/s)/(lb/in^2)$), then equations 8.1, 8.6, 8.2a, 8.3a and 8.4a apply directly (with $\theta_o$ replacing $y$). Substituting $\theta_i - \theta_o$ for $x$ in equation 8.4a gives the governing equation

$$\frac{K_q}{A}\,\theta_i = \frac{V_t}{4\beta}\frac{M}{A^2}\frac{d^3\theta_o}{dt^3} + \left(L + K_c + \frac{fV_t}{4\beta M}\right)\frac{M}{A^2}\frac{d^2\theta_o}{dt^2}$$
$$+ \left(1 + \frac{Lf}{A^2} + \frac{K_c f}{A^2}\right)\frac{d\theta_o}{dt} + \frac{K_q}{A}\,\theta_o \tag{8.7}$$

(for the four-way valve system of figure 8.4 with inertia and friction loading and with viscous leakage).

Each governing equation is a third-order linear differential equation with constant coefficients of the form

$$(a_0 D^3 + a_1 D^2 + a_2 D + a_3)\theta_o = a_3\theta_i \tag{8.8}$$

where $a_0$ equals the inverse of the square of the hydraulic frequency and $a_1$ is the damping term.

$$a_1 = \frac{2\zeta}{\omega_h} = K_c\,\frac{M}{A^2}$$

for pure inertia loading, but is given by

$$a_1 = \frac{2\zeta'}{\omega_h} = \left(K_c + L + \frac{fV_t}{4\beta M}\right)\frac{M}{A^2}$$

for inertia plus viscous friction and laminar leakage $a_2$ is unity but only approximately so for the more complex case, and $a_3$ is the velocity constant $(K_q/A)$.

A block diagram may be drawn for the system as in figure 8.3.

Equations such as 8.7 (or 8.8) are invaluable for analysis but the coefficients are not truly constant and any constant values assigned to them can only refer to small perturbations. This is a difficulty with hydraulic system linearised analyses. Non-linear methods with analogue computers can give more realistic modelling but analysis has the overriding advantages of illustrating general trends and of indicating suitable design values rapidly and simply.

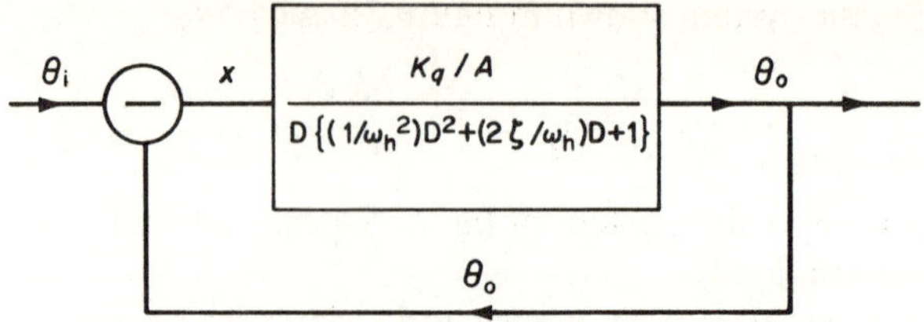

Figure 8.3 Block diagram for position control system

## 8.4  Feedback Lever System (with a Four-way Valve)

Another four-way valve position control system is shown in figure 8.4. The basic difference between this and the previously considered system is that there is not a 1:1 relation between input and output in the steady state. There is in fact a 'steady state gain' of $(\ell_1 + \ell_2)/\ell_1$. If the input were displaced by distance $N$, then, after transients had decayed, the output would have moved distance $N(\ell_1 + \ell_2)/\ell_1$.

The relations between valve spool displacements $x$ and load displacements (in this case $\theta_o$) can be derived in exactly the same way as those for the valve-controlled ram (see figure 8.1). The difference is that the loop is closed by means of the floating lever. For small motions, the distance between the pivot of the input rod and the other two pivots will be sensibly constant and equal to $\ell_1$ or $\ell_2$ respectively. This means that the governing equation is obtained by substituting $\theta_o$ for $y$ in equation 8.4 and replacing $x$ as

$$x = \frac{\ell_1 + \ell_2}{\ell_2}\,\theta_i - \frac{\ell_1}{\ell_2}\,\theta_o$$

The block diagram for the system of figure 8.4 can be presented as figure 8.5.

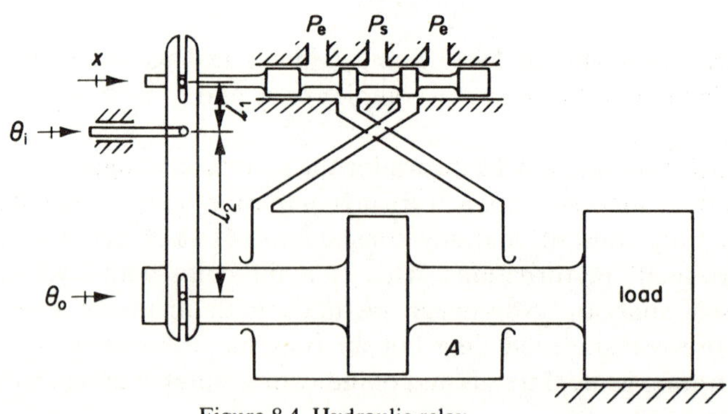

Figure 8.4 Hydraulic relay

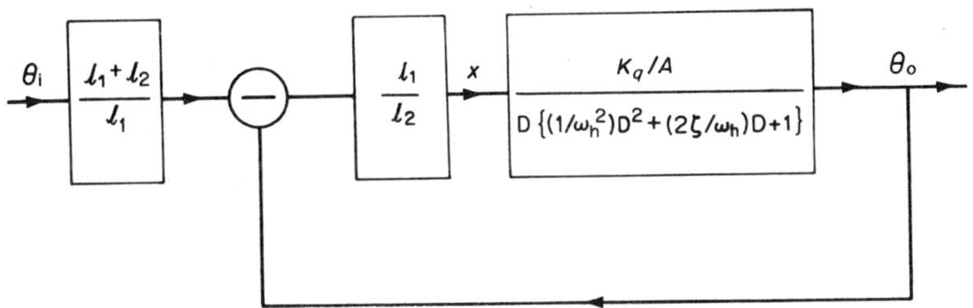

Figure 8.5  Block diagram for hydraulic relay

The governing equation is

$$\frac{K_q}{A}\frac{\ell_1+\ell_2}{\ell_1}\frac{\ell_1}{\ell_2}\theta_i = \frac{1}{\omega_h{}^2}\frac{d^3\theta_o}{dt^3} + \frac{2\zeta}{\omega_h}\frac{d^2\theta_o}{dt^2} + \frac{d\theta_o}{dt} + \frac{K_q}{A}\frac{\ell_1}{\ell_2}\theta_o \qquad (8.9)$$

(for the four-way valve system of figure 8.4 with feedback linkage) on

$$\frac{\theta_o}{\theta_i} = \frac{\ell_1+\ell_2}{\ell_1}\frac{(K_q/A)(\ell_1/\ell_2)}{(1/\omega_h{}^2)D^3 + (2\zeta/\omega_h)D^2 + D + (K_q/A)(\ell_1/\ell_2)} \qquad (8.10)$$

which is of the form

$$(a_0 D^3 + a_1 D^2 + a_2 D + a_3)\theta_o = a_3(k\theta_i) \qquad (8.8a)$$

where

$$a_0 = \frac{1}{\omega_h{}^2} = \frac{V_t M}{4\beta A^2}, \qquad a_2 \approx 1$$

(four-way valve)

$$a_1 = \frac{2\zeta}{\omega_h} = K_c\frac{M}{A^2} \text{ or } \left(K_c + L + \frac{f V_t}{4\beta M}\right)\frac{M}{A^2}$$

$$a_3 = \frac{K_q}{A}\frac{\ell_1}{\ell_2} \text{ and the steady state gain} = \frac{\ell_1+\ell_2}{\ell_1}$$

## 8.5  Valve Servo Characteristics

### 8.5.1  *Stability*

The first concern about a servomechanism is that it might be unstable when small disturbances (of the input, for example) would cause increasing oscillations of the load. The gain of the system must be limited. In the servos here considered this means limiting the supply pressure and the size of valve relative to the size of piston by limiting the value of the velocity coefficient

$(K_q/A)$. The acceptable limit will depend on the total mass and the oil compliance so the hydraulic frequency must be predicted realistically (excessive compliance due, for example, to air in the oil or additional mass due, for example, to long connecting lines would reduce this frequency). Damping is a major factor which can be increased by introducing viscous friction or shunt leakage. Predictions of damping factors $\zeta$ are likely to be inaccurate but the component of $\zeta$ attributable to the valve (the $K_c$ term) increases with valve opening and estimates are usually made for the worst conditions near the valve null where $K_c$ is low.

The analyses of the mechanical hydraulic servos in this chapter and in appendix B give rise to a third-order linear differential equation relating output displacement to input displacement, namely

$$(a_0 D^3 + a_1 D^2 + a_2 D + a_3)\theta_o = a_3\theta_i \tag{8.8}$$

The 'Routh criterion' for a servo system governed by equation 8.8 is that the system will be stable if $a_1 a_2$ is greater than $a_0 a_3$ and this criterion may be applied:

$$a_1 a_2 > a_0 a_3 \text{ means } \frac{2\zeta}{\omega_h} 1 > \frac{1}{\omega_h^2} \frac{K_q}{A}$$

giving the very useful general rule

$$\zeta > \frac{K_q/A}{2\omega_h} \tag{8.11}$$

(this is the criterion for the system shown in figure 8.2—the criterion would be

$$\zeta > \frac{K_q/A}{2\omega_h} \frac{\ell_1}{\ell_2}$$

for the system with a feedback lever of ratio $\ell_1/\ell_2$ as shown in figure 8.4).


### 8.5.2　Harmonic Response

The harmonic response of a control system is usually judged by assessing its behaviour under open loop conditions. In the case of the hydraulic servos just analysed, this means referring to their open loop counterparts, namely the valve-controlled system sketched in figure 8.1.

For an experimental harmonic test, the spool would be displaced sinusoidally (with amplitude $X$, for example) at different frequencies $\omega$ and the piston displacements measured. At high frequencies the piston would move very small distances from its midposition whereas, if the frequency were sufficiently low, the piston movements could be excessive (with the piston striking the ends of the cylinder). Within an appropriate frequency range both the amplitude of the piston displacements and the lag in phase of the piston relative to the spool would be found to vary with frequency. (An experimental problem is that pistons tend to drift from the central position owing to slight asymmetries but this is not considered here.)

Predictions* can be made of the piston displacements to be expected under harmonic testing for use in establishing the general characteristics of a proposed system although experimental results could differ in detail from these predictions.

The predictions would include a value for the amplitude ratio at the frequency for which the piston was 180° out of phase with the spool. Stability would be assured for the closed loop systems if this amplitude ratio was less than unity (for the system of figure 8.4 with a feedback lever, less than $\ell_1/\ell_2$). The 180° phase lag/unity amplitude ratio is the $(-1, i0)$ point on a Nyquist diagram.

The predictions would be made from the harmonic relation (transfer function) of the open loop system. This is obtained from equation 8.5 (or 8.5a or B.11)—the operational relation—after substituting $\theta_o$ for $y$ and $i\omega$ for D to give

$$\frac{\theta_o}{x} = \frac{K_q/A}{i\omega\{(i\omega/\omega_h)^2 + 2\zeta(i\omega/\omega_h) + 1\}} \tag{8.12}$$

### 8.5.3 *Description of Harmonic Response*

The behaviour of the open loop systems could be visualised by considering the above function as consisting of two terms.

The first term is $(K_q/A)/i\omega$ and it represents a pure integration. This would apply to an idealised valve piston combination with no load of any type and no shunt leakage. The mass of both piston and load would be zero, also the friction; and there would be no external forces—a situation which could only be approximated in practice. The pressures on the piston would be constant (at $P_s/2$) and the flow would depend only on $K_q$ and the spool displacement. Piston velocity would be proportional to spool displacement at any instant and piston displacement would thus depend only on the time integral of the spool displacement which implies (if $x = X \sin(\omega t)$)

$$y = \frac{K_q}{A}\int x\, \mathrm{d}t \quad \text{or} \quad y = \frac{K_q}{A}\frac{1}{\omega} X \cos(\omega t)$$

or

$$y = \frac{K_q}{A}\frac{1}{\omega} X \sin(\omega t - 90) \tag{8.13}$$

*For the validity of linear prediction see Royle (1959), Martin (1970) and Burrows (1972).

The output amplitude (that is the amplitude of the piston motion) would vary with the inverse of the frequency. A frequency of particular interest is $\omega = K_q/A$ for which the amplitude ratio would be unity. The output motion lags that of the input by 90° at all frequencies. A Bode plot is commonly used, viz. decibels versus octaves ($20 \log_{10}$ amplitude ratio being the logarithmic amplitude ratio in dB; and a doubling of $\omega$ representing an octave). The result above shows that doubling the frequency halves the amplitude ratio so for each increase of one octave frequency there would be a decrease of 6 dB ($20 \log_{10} 2 \approx 6$) in the amplitude ratio. The 0 dB point (unit amplitude ratio) occurs at $\omega = K_q/A$ and the Bode plot is a straight line as shown in figure 8.6.

The second term is

$$\frac{1}{(i\omega/\omega_h)^2 + 2\zeta(i\omega/\omega_h) + 1}$$

which has an exact analogy in a mass–spring–viscous damper system with

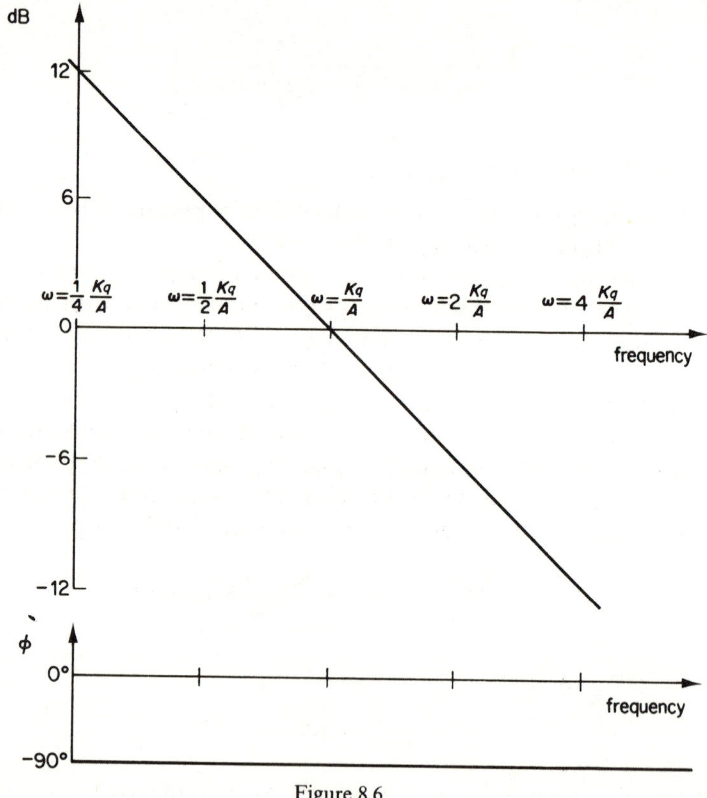


Figure 8.6

the same natural frequency as $\omega_h$ and the same damping factor $\zeta$. (It is based on there being only one mass—the spring and damper themselves having negligible mass—and a damping force directly proportional to velocity.) If the free end of the spring was moved sinusoidally, then the relation between movements of the mass and movements of the free end of the spring would be that given above. The displacement of the mass at any instant $t$ for a simple harmonic input displacement of unit amplitude and frequency $\omega$ would then be expressed by

$$\frac{1}{\{(1 - \imath^2)^2 + 4\zeta^2\imath^2\}^{1/2}} \sin(\omega t - \phi) \text{ for input } 1 \sin(\omega t)$$

(with the sine term replaced by the cosine for an input of $1 \cos(\omega t)$) where $\imath = \omega/\omega_h$ and $\tan \phi = 2\zeta\imath/(1 - \imath^2)$. For a very-low-frequency input, the output is of equal amplitude and in phase with the input. Alternatively at high frequency the output amplitude is small varying roughly with the inverse of the square of the frequency (Bode plot slope asymptotes to 12 dB per

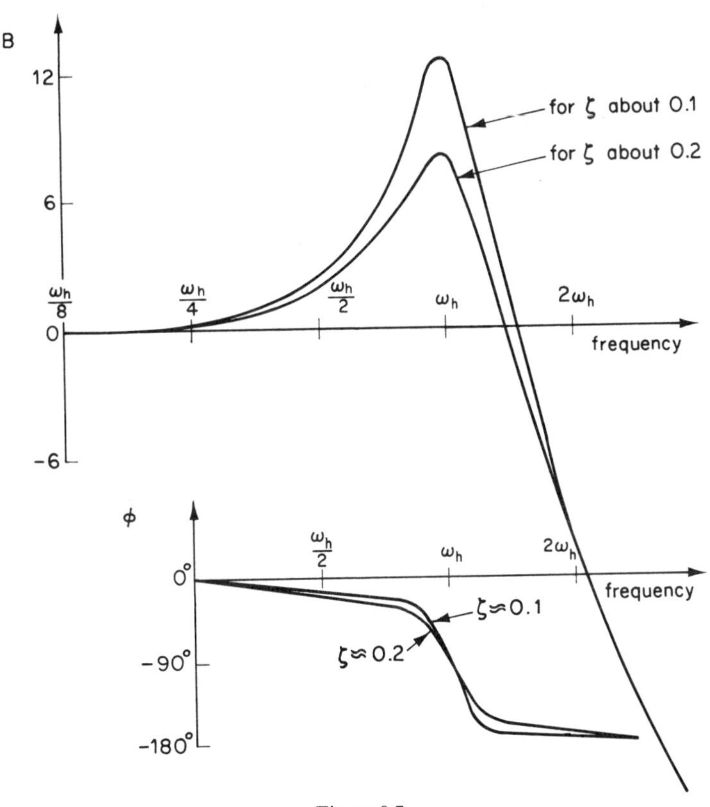


Figure 8.7

octave). At high frequency the phase angle $\phi$ approaches 180°. At the intermediate frequency of $\omega = \omega_h$, the amplitude ratio equals $1/2\zeta$ and the phase angle is 90°. Details of the Bode (logarithmic) plot vary with the damping ratio $\zeta$ and two examples are shown in figure 8.7, together with the related phase angles.

The two terms of equation 8.12 would apply to a combination of the two systems just considered where the output displacements of the (massless) piston (equation 8.13) became the input displacement of the free end of the spring. Physically this would not be practical as the force needed to move

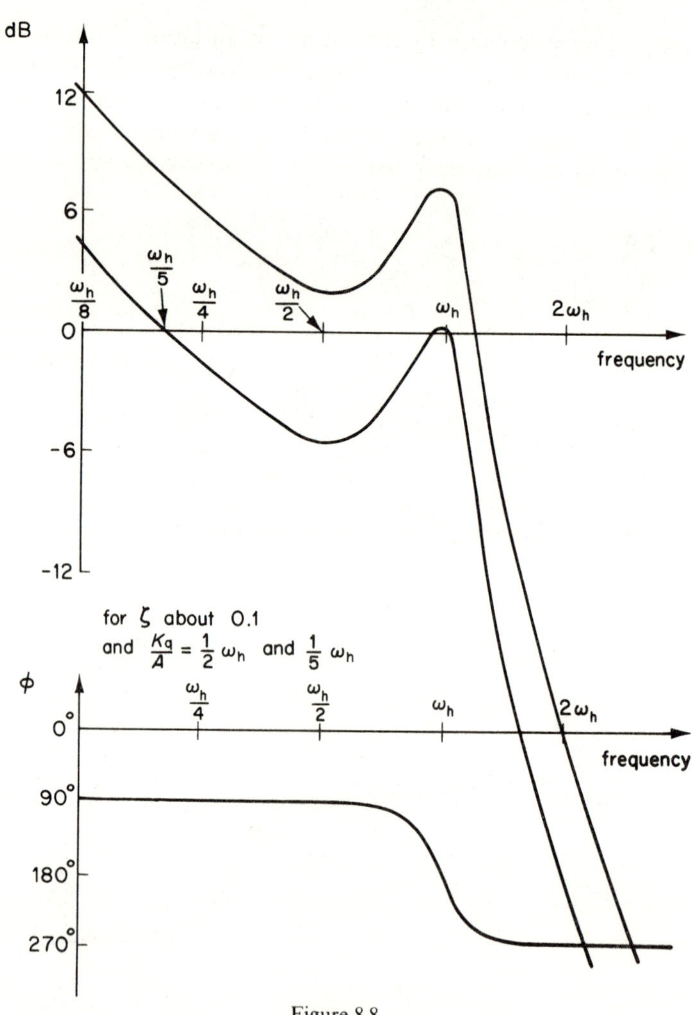


Figure 8.8

the spring would react on the piston. If it did happen, however, it would be a close representation of the hydraulic system. The phase *lag* of the complete system would be $90° + \phi$ and the two amplitude ratios would be multiplied together (or their logarithms added). Hence for a spool motion of $x = X \sin(\omega t)$ the output would be given by

$$\theta_0 = \frac{K_q}{A}\frac{1}{\omega} X \frac{1}{\{(1 - i^2)^2 + 4\zeta^2 i^2\}^{1/2}} \sin(\omega t - 90 - \phi)$$

and numerical values are most easily obtained by adding curves such as those in figures 8.6 and 8.7 graphically to give results such as those in figures

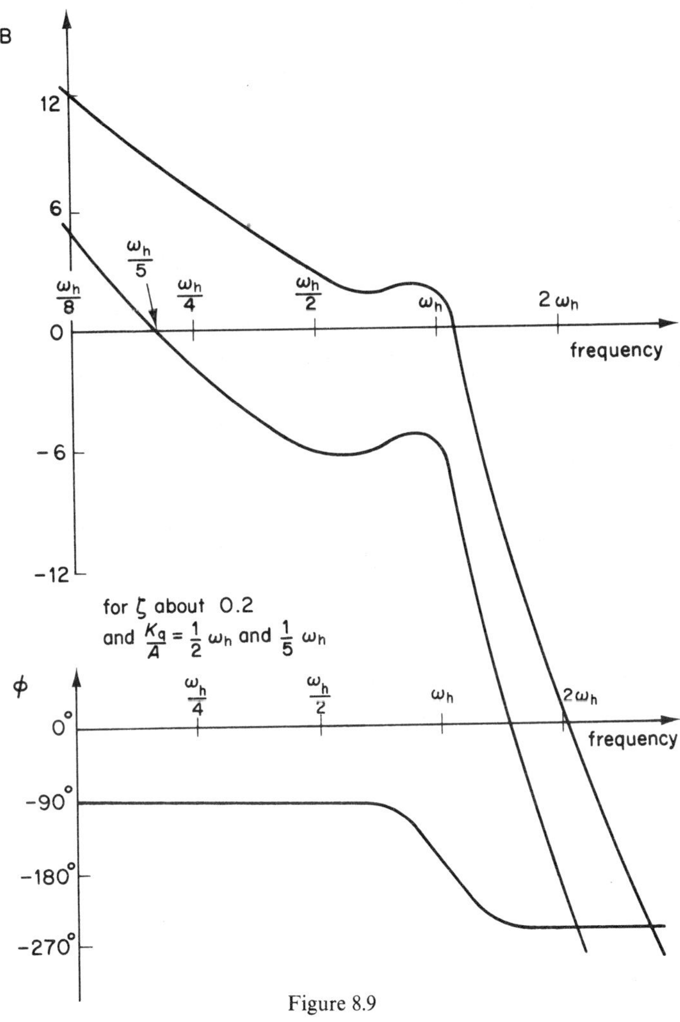



Figure 8.9

8.8 and 8.9 (at any frequency the two log amplitudes are added together as are the two phase angles).

### 8.5.4 *Open Loop Characteristics*

Sample amplitude ratio and phase angle predictions for the open loop system of figure 8.1 are shown in figures 8.8 and 8.9. The upper curves of both would refer to unstable systems (amplitude ratio greater than unity, that is positive logarithm, at the 180° phase angle). The lower curve of figure 8.8 would represent a marginally stable system (0 dB, that is unity amplitude ratio at 180°). The lower curve of figure 8.9 would refer to a stable system with a 'phase margin' of approximately 90° and a 'gain margin' of about 6 dB. Similar curves could, of course, be drawn for any damping ratio $\zeta$ and then moved vertically to correspond with any selected value of open loop gain $K_q/A$.

### 8.5.5 *Adequate Stability*

'Adequate' stability for a servo can only be defined empirically. The definition may be in terms of the gain and phase margins or of the maximum *closed* loop amplitude ratio or of maximum transient error (overshoot) following a step input or any combination of these.

Gain and phase margins may be read off open loop Bode plots just considered.

Maximum closed loop amplitude ratios may be predicted from numerical values for the open loop amplitude ratios and phase angles over a range of frequencies. One method is to draw the harmonic response locus (Nyquist diagram) and superimpose it on a 'Hall' (coaxial circles) chart. Another is to plot the numerical values with decibels as the ordinate and phase angles as abscissa and superimpose on a 'Nichols' chart. In both cases the maximum closed loop amplitude ratio can then be read off. A maximum amplitude ratio of 1.35 (that is 2.3 dB) is satisfactory for some applications.

Assuming that a specification is available defining the maximum acceptable amplitude ratio or the minimum acceptable gain and phase margins, the graphical technique described above would enable a suitable value of $K_q/A$ to be established if the system parameters $\omega_h$ and $\zeta$ were known. In practice the designer may be able to adjust all four parameters ($K_q, A, \omega_h$ and $\zeta$) in his search for a satisfactory design. One point of note is that $K_q$ varies with the square root of the supply pressure according to equation 7.6 so that the gain or velocity constant of a servo system changes with the operating pressure.

# 9 Electrohydraulic Servo Valves

Electrohydraulic systems use low-power electrical signals (of less than say 1 W) for precisely controlling the movements of large power hydraulic pistons and motors (which may be rated at say 10 hp—7460 W—or more). The 'interface' between the electrical (control) equipment and the hydraulic (power) equipment is the so-called 'electrohydraulic servo valve'. These valves are used on systems which must respond both quickly and accurately: aircraft controls are one example and numerically controlled machine tools another, although increasingly stringent specifications for other types of plant are extending their use into most fields. Many mechanisms which use other methods of control particularly if they already employ hydraulics could benefit from incorporating electrohydraulic techniques.

Assessing the suitability of an electrohydraulic servo valve for a particular application requires some insight into the features of different valve types which this chapter aims to give by explaining their modes of operation and 'steady state' behaviour and by including some information about their dynamic characteristics. At the end of this chapter are also some comments about electrical (control) supplies for such valves.

The dynamic analysis of the hydraulic components within a valve follows from the previous text and an outline of the dynamic features associated with electromagnetic actuators will be included here—a combination of both analyses would simulate a complete valve. However, it is rarely necessary to take full account of most valves' dynamic characteristics because their response is usually much more rapid than those of the systems they are used to control. This chapter is intended more to give sufficient background information for effectively using such valves than for designing them.

## 9.1  Flow Control Valves

Electrohydraulic flow control valves have a moving coil or moving iron device which positions a main control spool with a high degree of accuracy. The moving coil or moving iron component is called the 'armature' and small deflections of the armature cause displacements of the spool either directly by a mechanical link or indirectly via pilot pressure. The main spool can be of the three-way or four-way type and maximum spool displacements are commonly less than 2 mm. The widths of spool lands and valve body port lengths are typically matched to within a few micrometres whilst radial clearances between spool and bore may be as small as $5\mu$m.

Particles of 'dirt' in the fluid should be smaller than the radial clearance, for example smaller than 5 $\mu$m or 0.0002 in. Each port opening caused by movement of the spool acts as an orifice metering fluid to and from the load.

## 9.2 Valves with Coil Armatures

### 9.2.1 *A single-stage Valve*

The simplest type of electrohydraulic servo valve has a moving coil loud-speaker type assembly directly attached to a spring-centred spool valve as shown diagrammatically in figure 9.1. The coil is located in the field of a permanent magnet so that any current flow in the coil gives rise to a force in the axial direction. So long as the coil remains wholly in the magnet gap where the radial flux density is virtually constant, this force will be directly proportional to the current in both magnitude and direction.

The force applied by the coil will equal $kI_c$, where the constant $k$ is determined by the flux density $B_c$ in the gap, the coil diameter $d_c$ and the number $N_c$ of turns on the coil and, in the case illustrated, $k = B_c \pi d_c N_c$. (As a numerical example if $B_c = 1$ Wb/m$^2$, $d_c = 38$ mm and $N_c = 80$, the force equals 11.2 N/A, about $2\frac{1}{2}$ lbf/A.)

The magnitude and direction of this force (and of the current which causes it) will determine the movements of the spool. Neglecting any transient effects, the (steady state) spool displacement $x_{vs}$ for a constant coil current $I_{cs}$ will be given by

$$x_{vs} = \frac{k}{k'} I_{cs}$$

where $k'$ is the effective spring rate for the spool noting that $I_{cs}$ and hence $x_{vs}$ may be positive or negative.

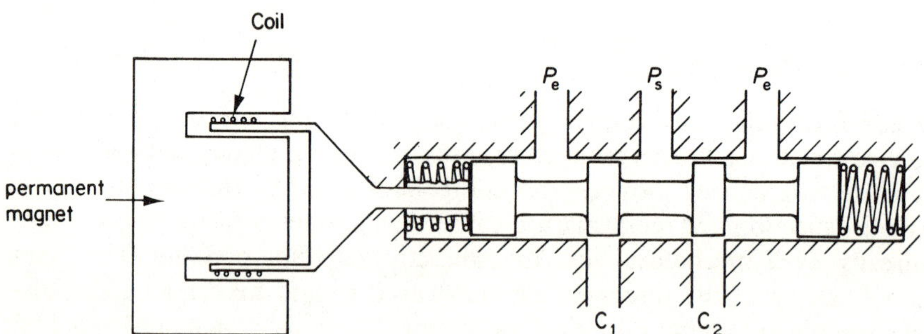


Figure 9.1 Diagrammatic representation of single-stage valve with moving coil armature

On this steady state basis, the spool displacement depends on the magnitude and direction of the current flowing in the coil (which in turn equals the voltage applied divided by the coil resistance). Different (steady) voltages or currents give different valve opening and, in so far as the spool spring rate $k'$ can be considered to have a constant value, the spool displacement or valve opening $x_{vs}$ under steady state conditions will be directly proportional to the coil current $I_{cs}$. We have shown already how valve openings and oil flowrates are related (in chapter 7 and appendix B) and, for the special case of a constant pressure drop across the ports of the valve, the oil flowrate is directly proportional to the valve opening. In this case the steady flowrate $q$ through the control ports $C_1$ and $C_2$ in figure 9.1 would be directly proportional to the control (coil) current or voltage.

When the voltage applied to the coil is changing, however, this simple directly proportional relation will not apply. One reason is that the current will not change immediately the voltage changes. Another reason is that the force applied by the coil has to overcome the dynamic forces of the spool, thus giving rise to an equation of the form

$$kI_c = m'\frac{d^2x_v}{dt^2} + f'\frac{dx_v}{dt} + k'x_v \tag{9.1}$$

and the relationship between coil *current* and spool displacement will be of second-order form or

$$\frac{x_v}{I_c} = \frac{k/k'}{(1/\omega_v^2)D^2 + (2\zeta/\omega_v)D + 1}$$

(where $\omega_v^2 = k'/m'$, $2\zeta/\omega_v = f'/k'$ and the 'effective' values of $m'$, $f'$ and $k'$ are considered in appendix A).

Electrically, the coil will have inductance $\mathscr{L}_c$ and resistance $R_c$ implying a first-order dynamic relation between the coil current $I_c$ and the voltage applied to the coil $e_{ap}$ given by

$$e_{ap} = \mathscr{L}_c\frac{dI_c}{dt} + R_cI_c \tag{9.2}$$

which is of the form

$$\frac{I_c}{e_{ap}} = \frac{1}{R_c}\frac{1}{1 + TD}$$

where $T = \mathscr{L}_c/R_c$ (the coil time constant).

Still another relation must be considered under dynamic conditions. The coil cuts lines of force at a rate which depends on its axial velocity $dx_v/dt$, thereby producing a back e.m.f. $e_b$ which acts against the applied voltage and equation 9.2 must be rewritten

$$e_{ap} - e_b = \mathcal{L}_c \frac{dI_c}{dt} + R_c I_c \tag{9.2a}$$

where $e_b$ in this particular case equals $B_c \pi d_c N_c (dx_v/dt)$ or $k(dx_v/dt))$.
Combining equations 9.1 and 9.2a

$$\frac{e_{ap}}{R_c} = T \frac{dI_c}{dt} + I_c + \frac{k}{R_c} \frac{dx_v}{dt}$$

$$\frac{k}{k'} \frac{e_{ap}}{R_c} = \frac{T}{\omega_v{}^2} \frac{d^3 x_v}{dt^3} + \frac{T 2\zeta}{\omega_v} \frac{d^2 x_v}{dt^2} + \cdots$$

$$+ T \frac{dx_v}{dt} + \frac{1}{\omega_v{}^2} \frac{d^2 x_v}{dt} + \cdots$$

$$+ \frac{2}{\omega_v} \frac{dx_v}{dt} + x_v + \frac{k^2}{k' R_c} \frac{dx_v}{dt}$$

note that the back e.m.f. acts as a damping effect on spool motion.

There is thus a third-order relation between *displacements of the spool* and *voltage applied* to the coil with a governing equation

$$a_0 D^3 x_v + a_1 D^2 x_v + a_2 D x_v + a_3 x_v = \frac{k}{k' R_c} e_{ap} \tag{9.3}$$

where

$$a_0 = \frac{T}{\omega_v{}^2}, \; a_1 = \frac{T 2\zeta}{\omega_v} + \frac{1}{\omega_v{}^2}$$

$$a_2 = T + \frac{2\zeta}{\omega_v} + \frac{k^2}{k' R_c}, \; a_3 = 1$$

$$k = B_c \pi d_c N_c, \; T = \frac{\mathcal{L}_c}{R_c}$$

and $k'$ is the effective spring rate for the spool.

### 9.2.2  *A Two-stage Valve*

Two-stage valves usually have an indirect link between the armature and the main control spool. Two-stage spool–spool valves have a first-stage spool valve directly attached to the armature. Figure 9.2 illustrates one possible configuration, the smaller (first-stage or pilot) spool acting as a three-way valve (see appendix B) controlling flow at the larger-area end of the main spool.

Referring to figure 9.2, a current applied to the coil causes a displacement of the pilot spool which—neglecting transient effects and possible variations

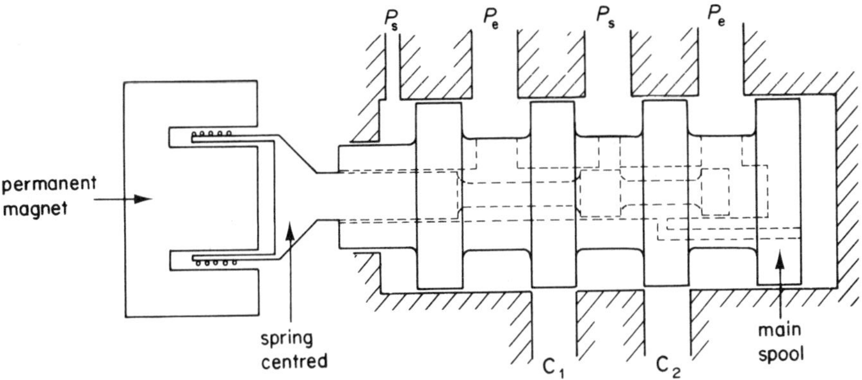


Figure 9.2 Diagrammatic representation of two-stage spool–spool valve

in the 'effective' spring rate due to oil flow — will be directly proportional to the current in both magnitude and direction. For example, a particular (steady) current may cause the pilot spool to move to the left by, say, $\frac{1}{2}$ mm. The effect of this movement would be to meter fluid from the supply into the larger-area end of the main spool, causing the main spool to move to the left. This movement would cause the pilot spool ports to close over the pilot spool lands with motion ceasing when the ports and lands were exactly aligned. At this point the main spool would have moved the same distance (in this example $\frac{1}{2}$ mm) and therefore the coil current would have caused the main spool to have moved by a distance which was directly proportional to that current. As with the single-stage valve, for the special case of a constant pressure drop across the main valve ports, the steady flowrate through the control ports $C_1$ and $C_2$ would be proportional to the steady coil current (or voltage).

Under dynamic conditions the directly proportional relationship between voltage, current and spool displacements would no longer apply. The two spools together represent a follow-up servo of the type shown in figure B.4 (and very similar to that of figure 8.2) with the main spool representing the load. The governing equation for this type of servo is the third-order equation B.12 (similar to equation 8.7 with minor differences in the coefficients) which relate displacements of the main spool (in equations B.12 and 8.7, $\theta_o$) to displacements of the pilot spool (in equations B.12 and 8.7, $\theta_i$). There is also another third-order equation relating displacements of the pilot spool to the coil voltage as for a single-stage valve; this is equation 9.3. If a full analysis of a two-stage valve is to be undertaken, then both these two equations (B.12 and 9.3) have to be incorporated but this is rarely necessary for the purposes of applying existing valves (see chapter 10).

## 9.3 Valves with Torque Motors

An important type of electrical actuator used in servo valves (mainly in two-stage valves) is the so-called 'torque motor'. A torque motor has an armature of magnetically permeable material whose small angular deflections actuate the first stage when control currents are applied. Torque motors often have two coils which may carry quiescent currents at the null position and changes in these currents modulate the flux in a magnetic circuit to displace the armature. Coils may be arranged with no quiescent current, the armature being displaced when current is supplied. The change of current in the first case or the actual current in the second case is termed the 'control' current $I_c$.

Two different types of torque motor are incorporated in figures 9.3 and 9.4 (which also show different types of first-stage and different operating modes). A permanent magnet type of torque motor is shown in figure 9.3 whilst a different type, requiring polarising current ('Law's relay'), is shown in figure 9.4.

The mode of operation of the type of valve shown in figure 9.3 may be briefly considered as follows. A control current in the coil(s) causes an angular deflection of the armature (clockwise, for example, as shown in the figure). Rigidly attached to the armature is the flapper of a nozzle–flapper valve

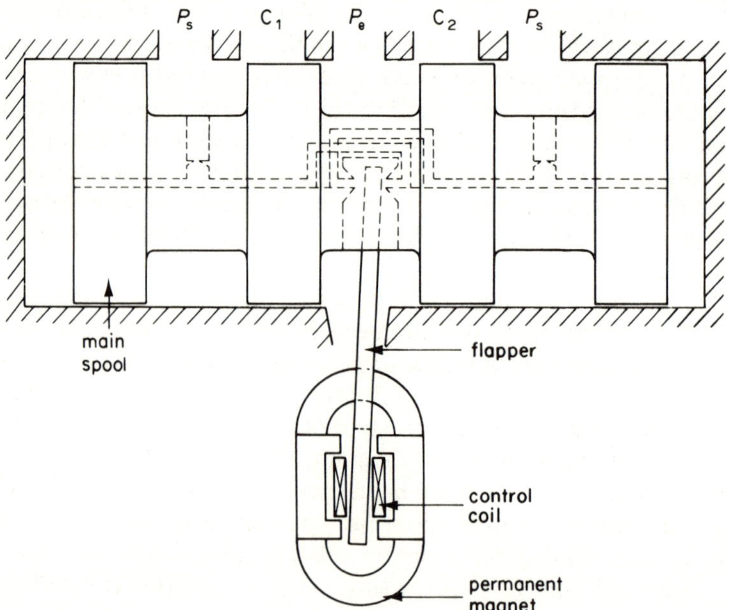


Figure 9.3 Diagrammatic representation of two-stage servo valve with permanent magnet torque motor, flapper–nozzle first stage and position feedback between stages

and the end of this moves in an approximately straight line (to the right as sketched, for example). The nozzles built into the main spool are affected by this flapper movement and the initially equal pressures on the two ends of the main spool become unbalanced. The unbalanced pressures cause the spool to move (to the right in the example sketched) until the flapper is again centralised between the two nozzles when motion ceases. In so far as the deflection of the armature can be considered directly proportional to the control current and the spool displacement to equal exactly the flapper displacement, the (steady state) displacement of the main spool will be directly proportional in magnitude and direction to the control current. The steady flowrate through the control ports would also be proportional to the control current if the pressure drop across the valve were constant.

The valve of figure 9.4 shows three main differences from that of figure 9.3. The first is that instead of a permanent magnet, it uses an electromagnet which requires a polarising current. For possible types of torque and force motor see chapter 11 of Blackburn *et al.* (1960). The second difference is that a jet pipe first-stage valve is employed instead of a nozzle–flapper. The third difference is that the main spool is not linked to the armature, so there is 'open loop' control over the main spool position relative to the armature. A similar type of 'open loop' valve but with a nozzle–flapper

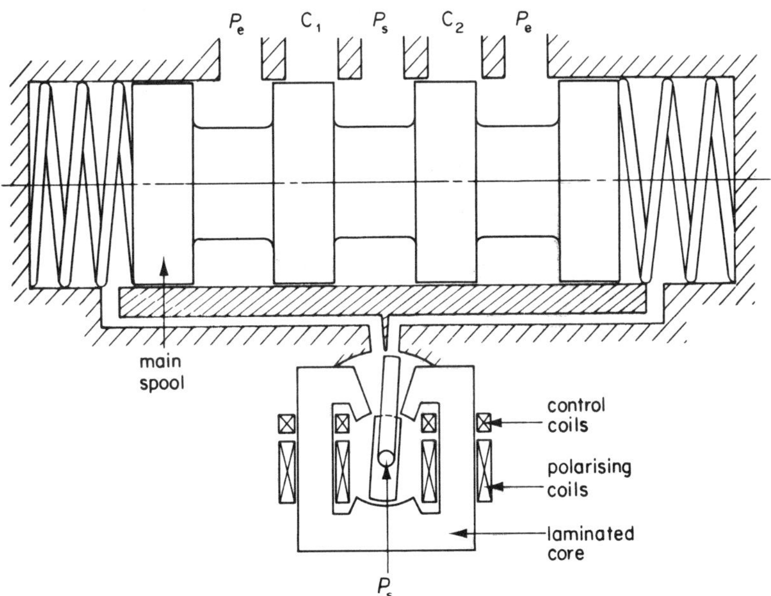


Figure 9.4 Diagrammatic representation of two-stage valve of the 'Law's relay' type with jet pipe first stage and open loop control of main spool

first stage is the subject of the paper by Nikiforuk *et al.* (1969). (Note that open or closed loop operation, sometimes with an electrical feedback link, jet or flapper or spool first stage, and various types of torque motor, may go together in any combination, see Shute and Turnbull (1961).)

The mode of operation of the valve shown in figure 9.4 is based on displacements of the armature directing the jet from the first-stage valve more onto the port connecting to one end of the main spool than onto the other. The effect of an armature displacement is to increase the pressure on one end of the main spool and to decrease the pressure on the opposite end. The spool then moves to compress the spring. The displacement of the spool is approximately proportional to the control current in the valve control coils under steady state conditions.

## 9.4 Valve Dynamics

The dynamic characteristics of two-stage torque motor valves will now be qualitatively considered. The torque motor relations could be combined with the relations developed in chapter 8 to give the overall characteristics. Simplified forms of valve characteristic representation will also be outlined.

### 9.4.1 *Torque Motors*

For the permanent magnet type of torque motor (illustrated in figure 9.4) a linear analysis is given by Merritt (1967) in terms of the magnetic constants and torque motor dimensions and of the forces acting on flapper valves. The force on either end of the armature, and hence the torque produced, is proportional to the difference between the squares of the magnetic fluxes in two opposing air gaps. Flux increases with the control current but also increases with the nearness of the armature to the appropriate pole piece (that is the armature has the tendency to snap onto the nearest pole piece once displaced from the null position). An approximate expression for the torque developed by the armature when displaced by angle $\gamma$ from the null position for a control current $I_c$ is

$$\tau_a = kI_c + k_m\gamma \tag{9.5}$$

Equation 9.5 incorporates the magnetic 'negative spring' with a rate of $k_m$ and for this type of valve a torsional centring spring must be provided of (positive) rate greater than $k_m$ for the device to operate—in practice this spring may appear in the form of a flexible mounting arrangement often termed a 'flexure tube'. The armature will be directly attached to the flapper of a nozzle–flapper valve or to the jet pipe of a jet pipe valve or linked to a pilot spool. The device will have inertia and some damping will exist so that the dynamic relation between control current $I_c$ and armature displacement $\gamma$ may be considered of the second-order form

$$kI_{c} = J_{a}\frac{d^{2}\gamma}{dt^{2}} + f_{a}\frac{d\gamma}{dt} + (k_{a} - k_{m})\gamma \tag{9.6}$$

(all coefficients referred to the armature noting that $k_{a}$ includes the centring spring and flow reaction spring effects).

Electrically, the coils will have self-inductance and (if two or more are used) mutual inductance so that, as equation 9.2, there will be a coil time constant $T(=\mathscr{L}_{total}/R_{c})$. Also any velocity of the armature will cause a back e.m.f. (proportional to $d\gamma/dt$).

The net effect (as for moving coil systems) is for there to be a third-order relation between displacements $\gamma$ of the armature or displacements $x_{f}$ of the first-stage valve and the applied voltage or

$$a_{0}D^{3}\gamma + a_{1}D^{2}\gamma + a_{2}D\gamma + a_{3}\gamma = Ke_{ap} \tag{9.7}$$

with the coefficients $a_{0}$, $a_{1}$, $a_{2}$ similar term by term to those of equation 9.3 incorporating the coil time constant $T$ and the armature natural frequency $(k_{a}/J_{a})^{1/2}$. The term $K$ is also similar involving the number of turns on the coil the magnetic flux in the air gaps and the torque motor dimensions (see Meritt (1967) for details). The equation 9.7 could alternatively be quoted for $x_{f}$ the variable where $x_{f} = r\gamma$.

The type of torque motor shown in figure 9.4 is less easily dealt with. However, it is self-centring so that a mechanical centring spring may not be needed but in other respects a similar dynamic relation can be taken for armature displacements versus applied voltage.


### 9.4.2 *Two-stage operation*

The two-stage torque motor operated valve of figure 9.3 has a 'follow-up' position control system built in because the main spool follows the position of the flapper. In the configuration shown in figure 9.3, the nozzle–flapper device acts as a four-way valve (for coefficients, see equations 7.12 and 7.13) and the main spool represents the load being controlled by this valve. Hence the assembly is similar to that shown in figure 8.2 and has the same type of governing equation, namely equation 8.7. This equation can be rewritten with $\theta_{i}$ representing the nozzle displacement (that is $\gamma$ times the radius arm, designated $x_{f}$) and $\theta_{o}$ representing the main spool displacement $x$ . This third-order equation contains terms relating to the hydraulic frequency and damping (of the main spool) whilst the velocity coefficient refers to the flow coefficient $K_{q}$ of the flapper valve coupled with the area $A$ of the ends of the main spool.

Combining equations 9.7 and 8.7 would give the sixth-order equation relating displacements of the spool and the applied voltage. It is not usually necessary to apply so complicated a relation when applying existing valves to control relatively low-frequency systems (see chapter 10) and even for valve design purposes some simplefication may be possible.

### 9.4.3 *Simplified Representations of Valve Characteristics*

9.4.3.1 *Analytical* Two simplifying assumptions could be made for analysing the behaviour of a two-stage valve. First the current build-up in the coil could be taken in the form of a simple exponential delay (that is first order relation) having the coil time constant $T$. This leaves the second-order relation between current and armature displacement although it may need some allowance for the damping effect of back e.m.f. Then the spool displacement could be taken as another simple exponential delay of time constant $T_2$ (equal to the inverse of the velocity constant for the first-stage–main spool assembly) by neglecting the dynamic forces on the spool (see analysis referring to figure 2.3). The result is a simplified relation between

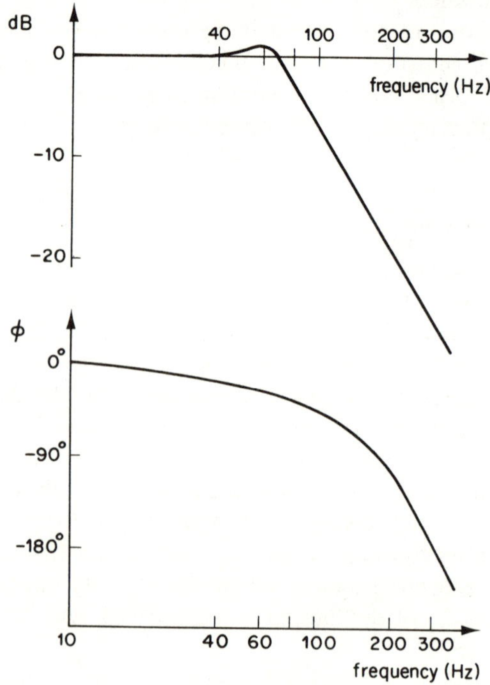


Figure 9.5

main spool displacements $x_v$ and applied voltage of the form

$$\frac{x_v}{e_{ap}} = \frac{K}{(1 + TD)\{(1/\omega_a^2)D^2 + (2\zeta/\omega_a)D + 1\}(1 + T_2D)}$$

*9.4.3.2 Semiempirical (experimental)* A practical alternative when dealing with an existing valve involves measuring its harmonic response over a range of frequencies. The type of result obtainable in this way is shown in figure 9.5, although this particular diagram is based on the coil current rather than the coil voltage.

The amplitude ratio for a valve, and its associated electrical circuitry at a particular test frequency, is obtained by knowing the amplitude of the main spool displacements when an a.c. voltage of given amplitude is being applied to the electrical amplifier which feeds or 'drives' the valve. This amplitude of spool displacement is divided by the displacement which occurs when a d.c. (steady and constant) voltage is being applied in the same way, the magnitude of this d.c. voltage being equal to the peak value of the a.c. voltage. In practice, such valve openings are difficult to measure and it is more usual to measure oil flowrates. For a given pressure drop across a valve the opening and flowrate are directly proportional. In practice it is also difficult to measure a rapidly changing flowrate such as that which occurs when an a.c. voltage is being applied to the electrical amplifier of a valve. However, it is the amplitude of such a flowrate divided by the steady flowrate for the same magnitude of input (d.c.) voltage which is used to give the amplitude ratio for the valve (normally expressed in dB). And the usual method of obtaining such results involves connecting the valve to a light-weight piston and electrically sensing the movements of the piston.

The phase angles for a valve are obtained by monitoring the input (a.c.) voltage and sensing the output flowrate variations. The lag of one sine wave relating to the other gives the phase angle. The amplitude ratios and phase angles for the valve are usually plotted for a range of frequencies on logarithmic axes.

These measured values may be approximated empirically to either a first-order or a second-order relation with appropriate time constant or frequency and damping ratio.

*9.4.3.3 Steady state approximation* A very simple analytical approach is often acceptable when dealing with complete servo systems. This consists of *completely neglecting the valve dynamics* and assuming the valve action to be instantaneous. The main spool displacement is then taken to be directly proportional to the control current at all instants or $x_v = K_1 I_c$, see chapter 10.

## 9.5 Comments on Electrical Supplies

In using the 'steady state' approximation it is usually more convenient

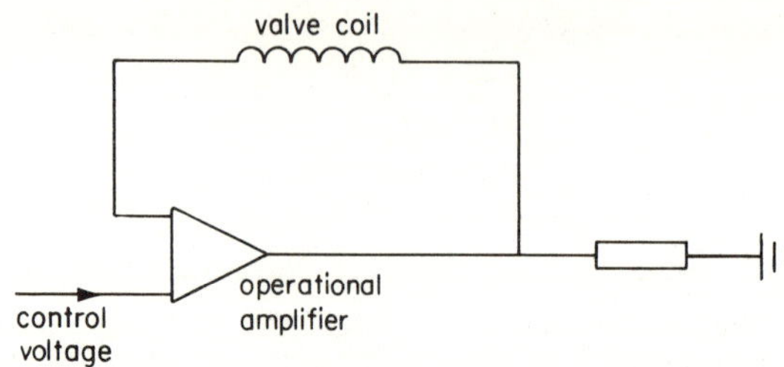


Figure 9.6 Diagram illustrating current forcing in valve coil

to take the valve opening or displacement as proportional to the applied
voltage (rather than the current). In some cases, however, the time constant
of the coil itself is not small enough to be neglected. (For example the coil
might have an inductance of 2 H and a resistance of 100 $\Omega$ involving a
time constant of 2/100 s (20 ms), the periodic time of the relatively low
frequency of 50 Hz.) To overcome this difficulty electrical methods are
available which effectively reduce the time constant. One technique involves
connecting the coil in the feedback loop of an operational amplifier as
illustrated in figure 9.6 instead of connecting it directly to the output.
The improvement obtained is related to the fact that an operational amplifier
acts to amplify any difference between the currents applied to it. If the
control voltage is fed to the amplifier through a pure resistance (equal
in magnitude to the coil resistance), the output voltage from the operational
amplifier will change rapidly whenever the current in the coil differs from
that in the input resistor.

Details of another method of effectively reducing the coil time constant
have been documented by Zeller (1968), with the claim that 'the electromagnetic torque motor coils, which have only a 40 Hz bandwidth when
driven by a conventional low output impedance amplifier' (see also Nikiforuk
*et al.* (1969)) 'exhibit flat response to 5000 Hz for 20% rated current and to
270 Hz for full rated current' (when driven by the amplifier described in
their publication).


### 9.5.1   *Pulse Width Modulation*

Whether or not special techniques are used to reduce effectively the coil
time constant, it is not always necessarily an analogue voltage which is
applied to a valve coil. An alternative technique is to provide the valve
input in the form of a train of pulses. For zero control signal, the alternate
pulses of positive and negative voltage are equal in duration and magnitude
as illustrated in the upper diagram of figure 9.7. The pulses keep the armature
oscillating through a small amplitude centred on the zero or null position.

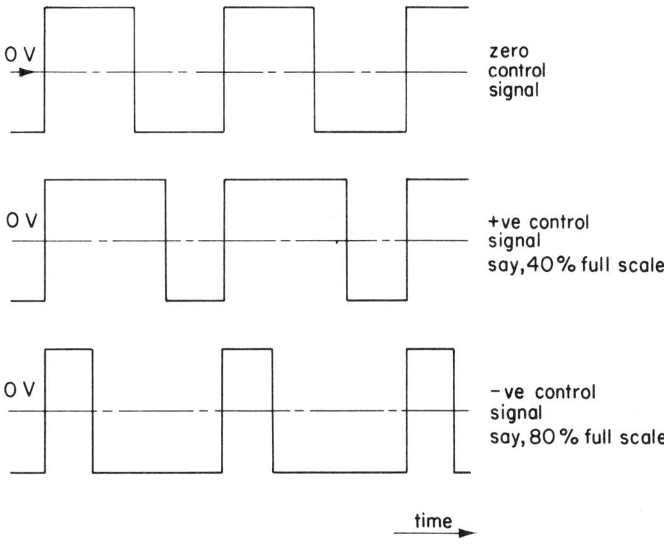



Figure 9.7

A control signal disturbs this symmetry, causing one pulse to last for a longer period and the other to last for a shorter period as indicated in figure 9.7. The effect is for the armature then to oscillate about a central point displaced from the null or zero by an amount which is proportional in magnitude and direction to the control signal. The technique is termed 'pulse width modulation' control or p.w.m. One advantage is the automatic provision of 'dither' on the valve which reduces the risk of the spool jamming in its bore (this is due, for example, to hydraulic lock, see appendix E). A suitable pulse frequency has to be found which can cause the valve spool always to be oscillating slightly without significant effect on the main hydraulic system.

# 10 Electrohydraulic Servomechanisms

Electrohydraulic servo systems combine together the versatility and precision available from electrical techniques of measurement and signal processing with the superior performance which high-pressure hydraulic mechanisms can provide when moving heavy loads and applying large forces.

Servos of this type are commonly used to operate the control surfaces of aircraft with actuators which are very compact because they operate at high pressure, for example 240 bar or 3500 lb/in$^2$ situated well away from the source of their electrical control signals. A different type of application is for automatic gauge control on sheet steel rolling mills. The servo system gives rapid modulation of the large forces in the mill housings in response to continuous (electrical) monitoring of the thickness of the sheet passing from the mill. Another steelworks' application is the cropping of rolled steel joists as the material leaves the final mill stand. A hydraulic piston traverses the saw alongside the moving material as cutting occurs, the motion of the piston being controlled by electrically sensing the position and speed of the moving joist. Another application is on numerically controlled machine tools with table motions actuated by electrohydraulic servos. In this case a hydraulic motor drives a lead screw under the control of electrical signals which represent respectively the demanded position of the table and the actual position at any instant.

The output of an electrohydraulic system may represent the variable which is actually being controlled as in the case of positioning a gun turret for example. However, the output may represent one link in a chain of control as, for example, when a hydraulic motor or piston is used to adjust a valve in a chemical plant.

Generally speaking with electrohydraulic servomechanisms, it is a position or a location which has to be controlled, although the versatility of electrical control devices and the relatively complex situations in which the servos are often employed tends to mask this simple fact. A good case can often be made for using such systems in simple applications—to replace devices of the on–off type which are, for example, often based on the use of solenoids—or to ease the manual operations in directly controlling large valves.

The aim of this chapter is to describe the basic principles of electrohydraulic position controllers and to give the essential linear analysis procedure relating to them.

Hydraulic systems can always be made to respond more quickly than electrical devices of the same power rating. However, in the case of electrohydraulic servomechanisms it is not the electrical features which normally limit their response. The electrical signal processing takes place almost instantaneously and occurs at a very low power level (for example less than $\frac{1}{2}$ W is involved in providing a 20 mA control current to a 1000 $\Omega$ valve coil). There is thus a rapid response even with large distances between the source of the control signals and the actual mechanism and, including the servo valve itself, it is usual for the electrical equipment to respond accurately to signals changing at rates of up to, say, 100 Hz. On the other hand, however, the hydraulic system may well be moving a large mass (of more than 1 tonne, for example) and its response must inevitably be relatively slow. Thus it is with electrohydraulic servos that the hydraulic components limit the performance of the system as a whole.

For analytical purposes, except where an electrical technique is purposely introduced for influencing dynamic behaviour, it will be assumed that the electrical response including that of the valve can be taken to be instantaneous. Electrical factors will then appear as constants in governing equations (that is with no dynamic terms). As a simple example a voltage which represents a displacement $\theta_o$ (in mm, for example) will appear as $K_e\theta_o$ (volts). The various electrical 'gains' of the system will then be incorporated into the hydraulic gain (that is the velocity constant) enabling the graphical (Bode diagram) technique to be applied again for devising systems which meet a given specification.

It is not possible to dissociate the electrical from the hydraulic features of an electrohydraulic system. The designer of the hydraulics, although needing access to electrical expertise for the details, must be in a position to specify at least the nature and magnitude of the necessary control signals.

Throughout this chapter, attention is concentrated on systems using four-way spool valves with nominally zero lap. The same methods could be applied with other types of valve.


## 10.1 'Proportional' Systems

Simple position control systems can be devised using a electrohydraulic servo valve for metering fluid directly to a piston or motor by applying to the electrical coils of the valve a voltage which is directly proportional to the error, that is to the difference between one voltage representing the desired or demanded position of the motor shaft or piston rod and a second voltage representing the actual position at any instant.

With a motor fed directly through a servo valve as sketched in figure 10.1,

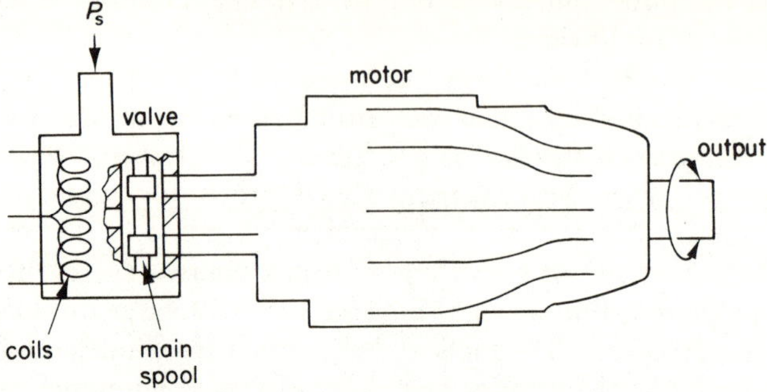



Figure 10.1  Electrohydraulic motor servo

the wiper of a rotary potentiometer could be attached or geared to the motor shaft so that the voltage of the wiper at any instant would represent the actual angular position of the shaft at that instant.

If a piston were used instead of a motor as illustrated in figure 10.2, a potentiometer wiper attached to the piston rod would sense rectilinear position in the same way.

For proportional control systems, this voltage representing actual position is subtracted electrically from the reference or demand voltage (which itself represents the demanded or desired position of the shaft or rod). Actually some operational amplifiers add voltages when the demand signal is then given the opposite polarity for convenience. The resulting difference between the two voltages represents the error and this voltage is applied to the coils of the valve (or some multiple of this error voltage depending on the voltage levels employed). In simple cases, the demand voltage could be set manually by simply adjusting another potentiometer for example.

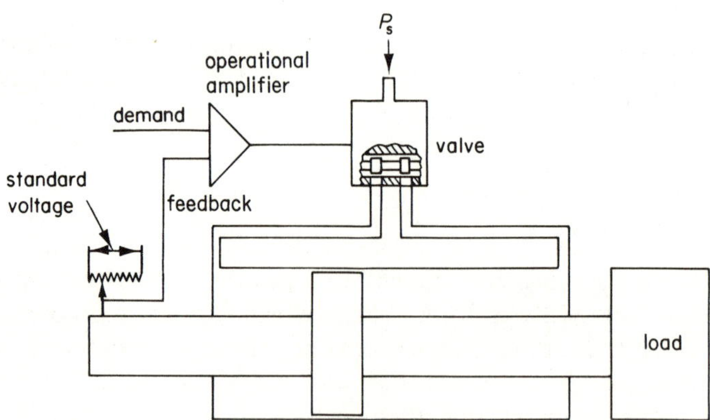



Figure 10.2  Electrohydraulic piston servo

In use and assuming the two voltages start off equal (so that the valve is closed and the system stationary), a sudden change in the demand voltage would initially cause a voltage to be applied to the coils, thereby causing the valve spool to be displaced. The resulting flow of oil through the valve would cause the piston or motor to move, thus altering the voltage from the output potentiometer wiper. The potentiometer voltage acting against the reference voltage (the feedback feature) would go to reduce the error voltage and thus to close down the valve. When the two voltages finally became equal again, the valve would shut and motion would cease. At this point the angular position of the motor (or linear position of the piston) should exactly equal the 'demanded' position, so providing a servomechanism which had 'zero position error'.

In practice, synchros or variable-position transformers or other devices are often preferred to potentiometers because of reliability. Such devices may involve modulating and demodulating circuits, for example, but no matter how complicated the electrical circuitry may become the essential requirement is for one d.c. voltage to represent the input or demand and another to represent the output. Their difference gives rise to a d.c. current (or possibly a pulse-width-modulated current) on the valve coils.

### 10.1.1 *Analysis*

As with any servo system, the stability and possible oscillation of these devices have to be assessed and the technique here is virtually the same whether a motor or a piston is employed. For the system with a double-acting piston (figure 10.2) the relation between piston displacements $\theta_o$ and valve spool displacements $x$ is that developed in chapter 8 for the system shown in figure 8.1, namely equation 8.5 which may be written

$$\frac{D\theta_o}{x} = \frac{K_q/A}{(1/\omega_h{}^2)D^2 + (2\zeta/\omega_h)D + 1} \tag{10.1}$$

(repeat of 8.5).

A valve-controlled motor (figure 10.1) has an exactly similar relation

$$\frac{D\theta_o}{x} = \frac{K_q/\delta_m}{(1/\omega_h{}^2)D^2 + (2\zeta/\omega_h)D + 1} \tag{10.1a}$$

The valve displacement $x$ in an electrohydraulic servo valve will be directly proportional to the control current flowing in the valve coil ($x = K_I I_c$)— if one neglects the dynamic characteristics of the valve considered in chapter 9—and the same equation still applies. (This is based on the assumption that the valve response will be an order of magnitude faster than that of the hydraulic system, see the end of this chapter for another possible approach.) Valve displacements $x$ may not be known explicitly but will be implicit in the rating of the valve—the rating of control current versus flowrate—and equation 10.1 is rewritten as

$$\frac{\mathrm{D}\theta_o}{I_c} = \frac{K_I K_q / A}{(1/\omega_h{}^2)\mathrm{D}^2 + (2\zeta/\omega_h)\mathrm{D} + 1}$$

(read $\delta_m$ for $A$ in a motor system) where $K_I$ is the (steady state) spool displacement per unit current (for example m/A or in/A). If the rating of a valve is known in terms of the flowrate for a given coil current and a given valve pressure drop, then the composite constant $K_I K_q$ can be calculated. As a numerical example for a valve to be used in a system operating at 210 bar (3046 lb/in$^2$) with data for the valve given as 0.5 l/s (30.5 in$^3$/s) for 20 mA control current at 70 bar (1015 lb/in$^2$) pressure drop, the value of $K_I K_q = 0.5 \times 10^{-3} \times (3)^{1/2}/0.02$ which equals $0.0433\,(\mathrm{m}^3/\mathrm{s})/\mathrm{A}$ $(2642\,(\mathrm{in}^3/\mathrm{s})/\mathrm{A})$.

The output displacement $\theta_o$ of the piston or motor will be measured as a voltage (conversion $K_e$ V/m or $K_e$ V/in for a piston and $K_e$ V/rad for a motor). Similarly the input or demanded position $\theta_i$ will occur as a voltage.

*Proportional* control will be assumed, meaning that the control signal will be a voltage directly proportional to the error and in this case equal equal to $K_e(\theta_i - \theta_o)$. This error voltage will usually be amplified and, assuming the amplifier has sufficiently fast response, this amplification can be represented as another constant $K_d$ (V/V). The coil current is the result of applying this amplified voltage to the valve coils, giving $K_R$ A/V if the coil resistance is $1/K_R$ and the coil time constant is negligible. The control current $I_c$ may be written as $I_c = K_R K_e K_d(\theta_i - \theta_o)$ so that the equation relating piston (or motor) speed to the error may be written

$$\frac{\mathrm{D}\theta_o}{\theta_i - \theta_o} = \frac{K_R K_e K_d K_I K_q / A}{(1/\omega_h{}^2)\mathrm{D}^2 + (2\zeta/\omega_h)\mathrm{D} + 1} \tag{10.2}$$

(read $\delta_m$ for $A$ in a motor system); (note equation refers to proportional control system) and a value of the 'velocity constant' $K_R K_e K_d K_I K_q / A = K_{qe}/A$ (with units of $(\mathrm{A/V}) \times (\mathrm{V/m}) \times (\mathrm{V/V}) \times (\mathrm{m/A}) \times (1/\mathrm{s})$, that is 1/s or s$^{-1}$) has to be selected. A similar decision is needed with mechanical servo systems where a suitable value of $K_q/A$ has to be found, *see* figures 8.8 and 8.9 reading $K_{qe}$ for $K_q$.

The velocity constant $K_{qe}/A$ should be chosen on the basis of the system damping factor $\zeta$ but this is not always calculable. Merritt (1967) suggests 0.1 to 0.2 as being reasonable characteristic damping factors for hydraulic servo systems which leads to the useful 'rule of thumb' that the system gain (that is the velocity constant) should not then exceed 20% to 40% of the hydraulic frequency of the system being controlled (see equation 8.11 reading $K_{qe}$ for $K_q$).

If a commercially available servo valve is used, then its 'rating' (in units of $(m^3/s)/mA$ has to be multiplied by $(P_s/\Delta P)^{1/2}$ to give $K_I K_q$, where $P_s$ is the supply pressure actually employed and $\Delta P$ is the total drop of pressure within the valve on which the maker's rating is based (often 69 or 207 bar, that is 1000 or 3000 lb/in$^2$). A valve would be chosen which gave enough flowrate for the maximum specified velocity of the system and $K_R$ (1/coil resistance) would be known for that valve. The valve selected should *not be too large* because that would accentuate the effects of minor imperfections in the valve.

With the value of $K_{qe}$ decided and the factors $K_I K_q K_R$ now known, it would only remain to determine the electrical gain $K_e K_d$ (probably in units V/m or V/mm, that is the voltage which would be applied to the valve coil if the error were 1 m (or 1 mm); or V/m (or V/mm) displacement of the piston).

As a numerical example, consider controlling the system referred to in chapter 3 having piston area $A$ $1.473 \times 10^{-3}$ m$^2$ (2.278 in$^2$) and hydraulic frequency 90 rad/s with a valve rated for $\Delta P = 70$ bar (1015 lb/in$^2$) at 0.05 (l/s)/mA (3.05 (in$^3$/s)/mA) having a 500 $\Omega$ coil. Assuming a supply pressure of 140 bar (2030 lb/in$^2$) the value of $K_I K_q$ is 70.7 (l/s)/A (4313 (in$^3$/s)/A). The choice of $K_{qe}/A$ is 20% to 40% of 90, say 30 s$^{-1}$. Now $K_{qe} = K_R K_I K_q K_d$ so that $K_e K_d$ in this case would be 0.312 V/mm travel (that is $1.473 \times 10^{-3} \times 30 \times 500/(70.7 \times 10^{-3}) = 312$ V/m or in British units 7.9 V/in (that is $2.278 \times 30 \times 500/4313$).

The selected velocity constant could be too large if the valve actually used had an unexpectedly underlapped spool assembly because the actual flow coefficient $K_q$ could have double the predicted value near its null position (compare equations 7.6 and 7.10). The extra damping which underlapping produces ($K_c$ increase, see equation 8.7 reading $K_{qe}$ for $K_q$ and equation 7.11) may not fully counterbalance such an increase in gain. The selected velocity constant may also be too large if the actual system has a lower than calculated hydraulic frequency due, for example, to (a) the presence of air, (b) the dilation of the cylinder or pipelines or (c) the extra mass effect of oil in the pipelines. The amplifier usually has a variable gain $K_d$ so that its contribution to the velocity constant can be adjusted.

### 10.1.2 *Precision*

These 'proportional' position control systems are type 1 servomechanisms giving precise positioning (nominally zero positional error) assuming there are no disturbing loads on the output or deflections of the structure in which the system is built. The presence of externally applied loads (or torques) would reduce the accuracy of control. For example, any steadily held load on a system when stationary (which could be static friction) necessitates a small opening of the control valve to sustain a pressure difference across the piston or motor for counterbalancing that load and to provide any leakage flow. An opening of the valve involves having a voltage on the coils which, for simple proportional controls, can only occur if some positional error exists.

## 10.2   Velocity Control

Simple velocity controls cannot satisfactorily be devised by simply using a tachogenerator to sense output speed and then applying to the valve the difference between this voltage (proportional to speed) and a demand or reference voltage. The reason this is not suitable is that the tachogenerator voltage can *never equal* the demand voltage (under steady state conditions), for then the valve would have to be shut. Although the output speed would bear some relation to the demanded value, control would be inaccurate and the system would not be a true closed loop servomechanism.

### 10.2.1   *Pump Control*

One satisfactory way of using an electrohydraulic servo valve for speed control involves metering fluid to a subsidiary cylinder which itself actuates the swash plate or eccentric device of a variable-delivery pump. For speed control, with a tachogenerator providing a voltage proportional to the speed of a fixed-displacement motor supplied by the pump, it then becomes possible to operate with zero 'velocity error' (figure 10.3). For example, a change in the reference (demand) voltage would cause the valve to open, providing oil to operate the ram attached to the pump control. As pump flow increased, say, the tachogenerator (feedback) voltage would increase and when it attained the same value as the reference voltage the valve would

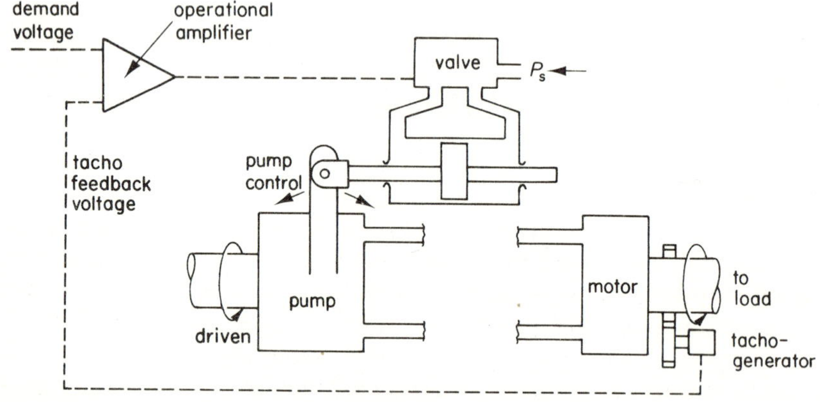


Figure 10.3 Electrohydraulic variable pump servo

shut. In this case there need be no difference between reference and feedback voltages for a constant motor speed to be maintained which makes possible a zero 'velocity error' system.

This essential improvement occurs because an extra 'integration' has been introduced in the loop. The pump control ram will move at a *speed* which is (approximately) proportional to valve *opening* (or displacement). Thus the *displacement* of the pump controller is proportional to the time *integral* of the servo valve displacement.


## 10.3   Compensated Control

'Compensated' valve control systems can be devised, to exploit the advantages of both 'proportional' and 'integral' control. Such systems can give accurate position control or accurate speed control even in the presence of disturbing loads (or torques).

The principle is to feed to the valve coils a voltage which is the sum of two voltages, the first proportional to the error and the second proportional to the time integral of the error.

As a notional example for a position control system consider a forklift truck which has to raise different weights to precise heights—operated perhaps by a piston fitted with an inline potentiometer to sense the height of lift. With a heavy load, the desired position would be approached but the valve would have to remain slightly open (that is with some current on the valve coils) to sustain pressure on the underside of the ram (see equation 7.1 for $q = 0$). Using simple proportional control, this would involve an error or difference between the actual and demanded positions (represented as a difference between the potentiometer and reference voltages). Using some degree of integral control, however, an error existing for any significant length of time would begin to build up as relatively large voltage applied to the valve coils (owing to the integration with respect to time). Thus the valve would operate to raise the load further until eventually there was zero error despite a sustained voltage (and hence coil current) keeping the valve partly open. On the other hand, however, with a light load to be lifted, the integral effect might build up excessively whilst the lift was occurring causing an overshoot. Afterwards, the error (in the opposite sense) would begin to integrate and open the valve the opposite way until finally settling. Integral control can cater for external loading but gives more oscillatory response and the size of the integration signal must be limited (by selecting an appropriate integration time constant).


### 10.3.1   *Analysis*

To introduce the 'compensated' (integral plus proportional) signal to an otherwise simple control system such as that shown in figure 10.2 implies altering the operational relation of equation 10.2 to the form

$$\frac{D\theta_o}{\theta} = \frac{(1 + 1/T_iD)K_{qe}/A}{(1/\omega_h{}^2)D^2 + (2\zeta/\omega_h)D + 1} \qquad (10.3)$$

(where $K_{qe} = K_R K_e K_d K_I K_q$ and $\theta$ is the (positional) error namely $\theta_i - \theta_o$: also note in the case of a motor system read $\delta_m$ for $A$).

Although the integration feature $1/T_iD$ is needed to improve steady state accuracy, the magnitude of $T_i$ must be chosen with due regard to its effect on dynamic characteristics.

### 10.3.2  *A Possible Electrical Network*

A well-known electrical network for use in integral control, shown in figure 10.4, gives an approximation to the required operational relation $1 + 1/T_iD$. However, as recommended by Merritt (1967), the capacitor should be bridged by biased diodes to avoid excessive positive or negative voltages. Without such biasing if a large error existed for even a short length of time, its time integral could exceed the electrical saturation voltage and could drive the spool valve to its fully open position.

The input to the network is the error signal $e_1$ and the output $e_2$ is the control signal which goes on to be amplified before being supplied to the valve coils. The relation between $e_2$ and $e_1$ is

$$\frac{e_2}{e_1} = \frac{1 + (1/\alpha)T_iD}{1 + T_iD}$$

(where $T_i = RC$) a function which approximates to 1 if $e_1$ is changing slowly but which (for $\alpha > 1$) approximates to

$$\frac{1}{\alpha}\left(1 + \frac{1}{T_iD}\right)$$

If $e_1$ is changing rapidly (for example sinusoidally at a frequency which is

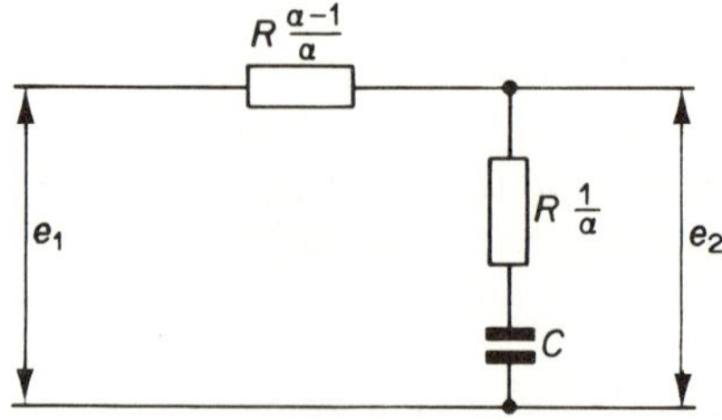


Figure 10.4  Compensating network

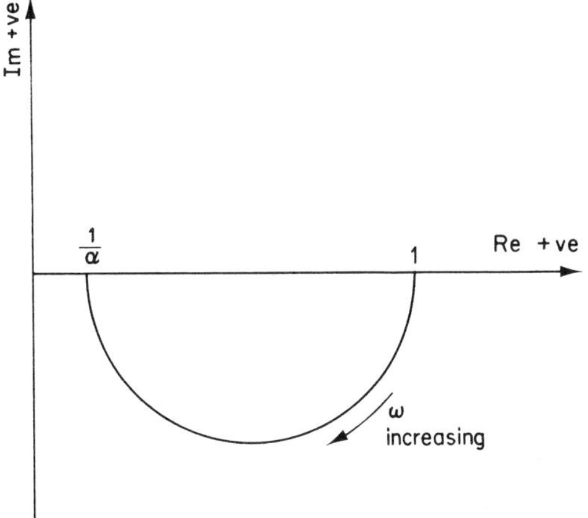

Figure 10.5  Locus for network ($e_1$ is represented as the unit vector)

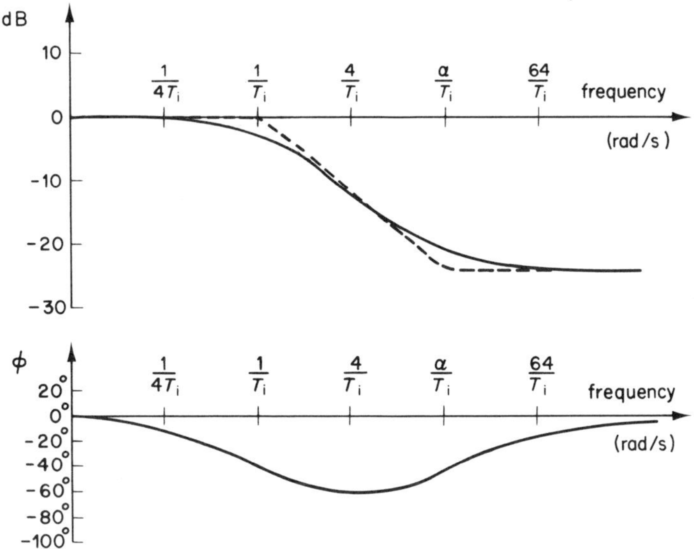

Figure 10.6  Logarithmic plots for network ($\alpha = 16$)

high compared with $1/T_i$). The harmonic response locus for this network—
the plot of the function

$$\frac{1 + i\omega(1/\alpha)T_i}{1 + i\omega T_i}$$

—is shown in figure 10.5 and on logarithmic axes in figure 10.6. The value of $\alpha$ normally lies between 10 and 20.

The fact that the network is attenuating (the amplitude of $e_2$ tends towards $1/\alpha$ times the amplitude of $e_1$ at high frequency) enables an increase to be made in the velocity constant $K_{qe}/A$ of any electrohydraulic system in which the network is incorporated, most probably by increasing the gain $K_d$ of the electrical amplifier about $\alpha$ times.

The value of $T_i$ has to be selected to suit the system with which the network is to be used. For rapid corrective actions, the integrating time constant

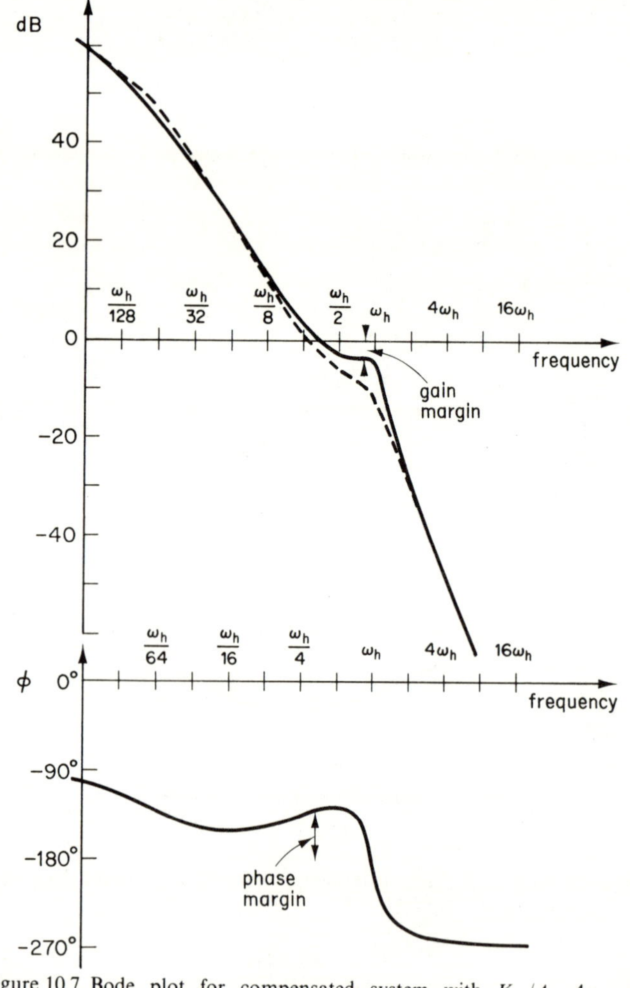


Figure 10.7 Bode plot for compensated system with $K_{qe}/A = 4\omega_h$ and $T_i = 64/\omega_h$

$T_i$ should be short. However, it must not be so short as to worsen significantly the stability of the complete servo. The problem of selecting a suitable value for $T_i$, whilst retaining the advantage of the increased velocity constant, may be resolved graphically.

### 10.3.3  *Systems with a Compensating Network*

The graphical method of seeking a suitable value for $T_i$ involves modifying the open loop Bode diagram of the main electrohydraulic system to incorporate that of the network. A system diagram such as the lower curve of figure 8.9 (reading $K_{qe}$ for $K_q$) would be obtained for a simple (proportional control) system. The network is attenuating and the gain of the drive amplifier would be increased ($\alpha$ times) so raising the system amplitude curve by $20 \log_{10} \alpha$ (say $20 \log_{10} 16$) without affecting the phase angles. Then the amplitude values for the network would be added (all values are zero or negative in fact) to give the final amplitude plot for the compensated system. Similarly the phase angles for the network would be added (again zero or negative) to those for the simple system to give the phase angle plot for the compensated system. The type of result obtained by combining the logarithmic plots of the system and the network is illustrated in figure 10.7.

The dynamic characteristics of the compensated system must be satisfactory as assessed in terms of the gain and phase margins or otherwise.

This integration feature is often termed phase lag compensation. If a flow control valve is being used to control position, this form of compensation helps reduce error due to external loads. If a flow control valve is being used to control speed, then this (or a similar) form of compensation is virtually essential.

## 10.4  Valve Characteristics

It has been assumed that the response of any electrohydraulic servo valve will be an order of magnitude faster than that of any hydraulic system which it is being used to control. This will not always be the case as the valve may be used to control a particularly stiff hydraulic system (with high hydraulic frequency) or the valve itself may have a slow response. The original assumption implies that the amplitude ratio and phase angle for the valve will be 0 dB and 0° respectively at all frequencies which are of relevance as far as the hydraulic system is concerned. Extending this to include all frequencies is the basis of applying figures 8.8 and 8.9 directly for electrohydraulic servos of the 'proportional' type.

One method of incorporating the dynamic characteristics of a valve is to modify the Bode diagram of the basic system by adding in the amplitude and phase values for the particular valve being used. The outcome in many cases would not be worth the effort involved because the amplitude and phase for many valves only begin to differ from 0 dB and 0° above about 50 Hz whereas many systems have hydraulic frequencies much below this.

If it is decided to assess the effect of the valve's dynamic characteristics on the behaviour of the system, it is first necessary to obtain the data for the valve preferably in the form of an (open loop) logarithmic diagram. The diagram should relate to the valve complete with its electrical circuitry because the characteristics will depend not only on the hydraulic and magnetic/electrical characteristics but also on both the behaviour of the electrical amplifier and the way it is employed (for example see figure 9.6). However, any compensating circuit may well be dealt with independently.

To assess the effect on the dynamic characteristics of a servo system, the procedure is to add the (log) amplitude values for the valve to those for the system at the same frequencies in order to obtain the amplitude plot for the whole device. Phase angles are similarly dealt with.

For making up a Bode diagram which incorporates the valve dynamic characteristics it is possible to use the manufacturer's data for the valve but care must be taken to ensure that the characteristics of the electrical amplifier and circuitry are also included.

# 11 Conclusion

The practical development work needed to produce a satisfactory high-performance hydraulic system will by no means be eliminated when the system has been designed by using the methods of linear analysis described in this book. The same would be true even if more complicated methods of calculation were adopted. It is doubtful whether any such system has ever been developed from slide rule and drawing board alone. It is equally doubtful whether a system could be produced by trial and error methods alone. A satisfactory system will probably always need both practical and analytical treatment.

Some practical ruses to improve troublesome systems such as the judicious polishing of a valve spool to prevent jamming or the changing of a pipelength to eliminate an undesirable vibration are qualitatively suggested in this text (appendix E and figure 1.4 respectively). However, many practical weaknesses are less easily recognised such as uneven valve laps or variable seal friction forces or (in the case of electrohydraulic systems) electrical noise and there are many factors influencing the detailed characteristics of a system which have not been dealt with here.

The method of linear analysis is probably the most convenient and simplest method currently available for system design calculations. Any simpler method would be in danger of giving misleading results. More complicated methods may sometimes be needed to refine certain predictions but in any case the linear method would be tried first.

Perhaps the main advantages of using linear methods for design calculations are the relative ease with which predictions can be made—linear analyses are not particularly time consuming—the way in which these methods give the designer the 'feel' of a system and finally the significant reduction in development effort which comes from doing such calculations before the practical work begins.

# Appendix A   Spool Valve Stroking Forces

The axial force needed to slide a valve spool in its bore can conveniently be split into three components: (1) the force needed to accelerate the mass of the spool and anything moving with it, (2) the force needed to overcome friction and (3) forces acting on the spool due to flow of fluid through the valve. There may also be another force due to biasing or centring springs. We shall begin with flow forces following the method of Lee and Blackburn.

## A.1  Flow Forces

### A.1.1  *Steady Flow*

Consider a simple valve with fluid leaving through a very small port opening (that is $x \ll d_1$) as indicated in figure A.1.

A jet (of annular shape) forms at the outlet and the volume flowrate may be written

$$q = C_{\mathrm{d}}\pi d_1 x (P_{\mathrm{a}} - P_{\mathrm{b}})^{1/2}(2/\rho)^{1/2} \qquad (A.1)$$

(see also equation 1.2).

An exit angle $\theta$ can be assigned to the jet. This is usually taken to be 69°.

The jet causes a reaction force (as in the case of a jet issuing forth from a rocket) which equals the rate of change of momentum of the fluid. In the

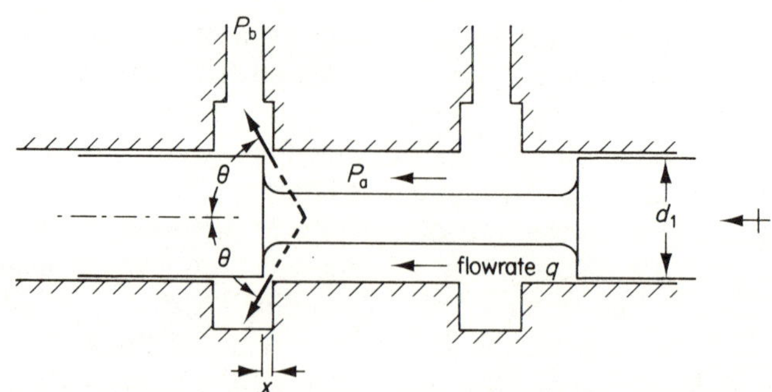


Figure A.1

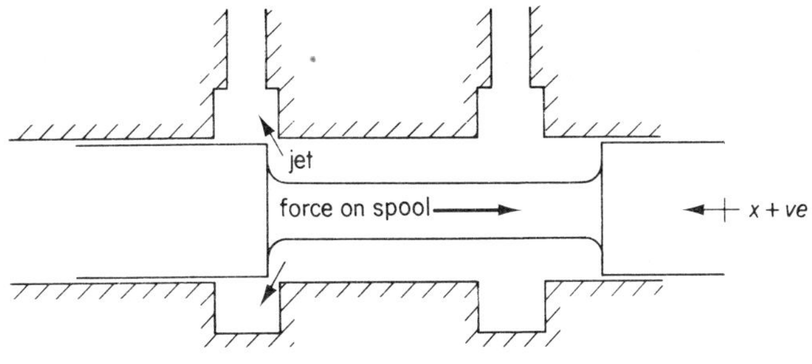



Figure A.2

valve, the fluid before the orifice is assumed to have zero velocity and hence
zero momentum, so the force due to the jet can be written

$$F_{\text{jet}} = q\rho \, \frac{q}{C_{\text{d}}\pi d_1 x} \qquad (A.2)$$

(where $q/C_{\text{d}}\pi d_1 x$ is taken to be the representative jet velocity).

The axial component of this momentum force, namely $F_{\text{jet}} \cos\theta$ appears
as a force acting on the spool as indicated in figure A.2 (there being reduced
pressure on the annular face near the jet).

Thus there is a 'steady flow force' acting on a spool in such a direction
as to close the valve, which has magnitude

$$F_{\text{mom}} = C_{\text{d}}\pi d_1 x(P_{\text{a}} - P_{\text{b}})^{1/2}(2/\rho)^{1/2}\rho(P_{\text{a}} - P_{\text{b}})(2/\rho)^{1/2}\cos\theta$$

or

$$F_{\text{mom}} = 2C_{\text{d}}\pi d_1 x(P_{\text{a}} - P_{\text{b}})\cos\theta$$

or (substituting $C_{\text{d}} \approx \tfrac{5}{8}$ and $\theta \approx 69°$)

$$F_{\text{mom}} \approx \tfrac{7}{16}\pi d_1 x(P_{\text{a}} - P_{\text{b}}) \qquad (A.3)$$

The steady flow force is similar to a spring force (being directly proportional to $x$ for a given pressure drop across the orifice). In calculating forces
on a spool one can term it the 'Bernoulli spring force'. A numerical example
for a valve of the type shown in figure A.1 with spool diameter $d_1 = 10$ mm
(0.394 in), pressure drop across orifice 35 bar (507 lb/in$^2$) and port opening
$x = 0.3$ mm (0.012 in) gives

$$F_{\text{mom}} \approx \tfrac{7}{16} \times \pi \times 10 \times 10^{-3} \times 0.3 \times 10^{-3} \times 35 \times 10^5 \approx 14.4\,\text{N}$$

(or in British units $\tfrac{7}{16} \times \pi \times 0.394 \times 0.012 \times 507 \approx 3.3$ lbf)—with similarly
28.8 N for 0.6 mm opening. The flowrate for 0.3 mm is (see equation 1.2a)
approximately

$$10 \times \pi \times 10 \times 10^{-3} \times 0.3 \times 10^{-3} \times (35)^{1/2} = 0.55 \times 10^{-3}\,\text{m}^3/\text{s, that is } 0.55\,\text{l/s}$$

(or in British units $100 \times \pi \times 0.394 \times 0.012 \times (507.6)^{1/2} = 33.5$ in³/s). Note that the Bernoulli 'spring rate' is 14.4/0.3 or 48.1 N/mm (or 270 lbf/in).

Reversing the direction of flow through a valve does *not* reverse the direction of the momentum force. The momentum change of an entering jet gives a force on the spool in the same direction as the flow. Referring to figure A.2, if the flow enters (rather than leaves) as a jet at the annular orifice, the force on the spool still acts in the direction shown.

### A.1.2   Transient Flow

A changing flowrate through the valve gives rise to another force component. Referring to figure A.3, there is fluid in the valve chamber of approximate mass $\rho\ell(d_1{}^2 - d_2{}^2)\pi/4$.

The flow is taken to be increasing at rate $\dot{q}$ and the average fluid velocity in the chamber at rate $\dot{q}/(d_1{}^2 - d_2{}^2)(\pi/4)$, which is the rate of acceleration of the mass of fluid in the chamber. The force causing this acceleration requires an equal and opposite reaction on the spool. Hence the force on the spool due to flow acceleration acts in the opposite sense to the direction of the acceleration, and is given by

$$F_{\text{acc}} = \dot{q}\rho\ell \tag{A.4}$$

The rate $\dot{q}$ of change of flowrate may be expressed (see equation A.1) as follows if both $x$ and $P_a - P_b$ are variable:

$$\dot{q} = \frac{\mathrm{d}q}{\mathrm{d}t} = C_d\pi d_1(2/\rho)^{1/2}\left\{(P_a - P_b)^{1/2}\frac{\mathrm{d}x}{\mathrm{d}t} + \frac{x}{2(P_a - P_b)^{1/2}}\frac{\mathrm{d}(P_a - P_b)}{\mathrm{d}t}\right\}$$

or alternatively if $x$ varies but $P_a - P_b$ can be considered constant as

$$q = C_d\pi d_1(P_a - P_b)^{1/2}(2/\rho)^{1/2}\dot{x}$$

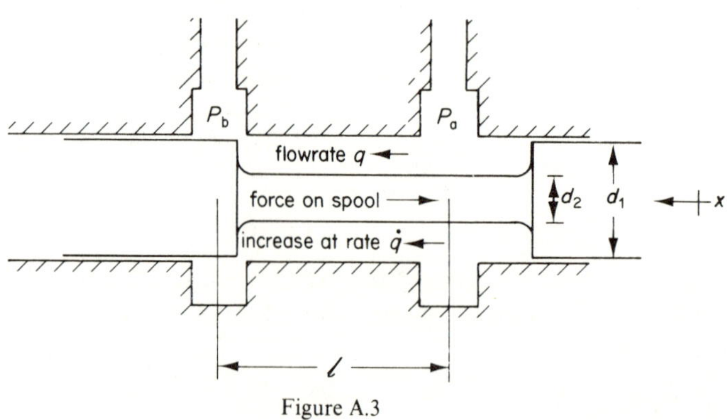


Figure A.3

showing that the rate of change of flow may be considered proportional to spool velocity.

Thus the transient flow force is similar to a damping (viscous friction) force on the motion of a spool.

Reversal of the flow direction does reverse the direction of the force $F_{\text{acc}}$ and then the force may have a destabilising effect analogous to 'negative damping'.

## A.2 Effective Moving Mass

The mass of metal comprising the spool obviously has to be accelerated and decelerated whenever the spool changes position, and the mass of any solid parts attached to it, also one-third of the mass of each spring acting directly on the spool. In addition, however, the acceleration of a spool usually involves the acceleration of some fluid as well.

Consider a valve as sketched in figure A.4, moved mechanically and with drain ports to vent oil from the ends of the spool. The spool has mass $m_1$, the valve chambers contain a mass of oil $m_2$, and each drain line, running full, contains a mass of oil $m_3$.

In the simplest analysis, the total mass to be accelerated is $m_1 + m_2 + 2m_3$ of which the last may be dominant. For example, a 6 mm (0.24 in) diameter spool of length 20 mm (0.79 in) might give $m_1 + m_2 \approx 0.0045$ kg (0.01 lb) whereas 6 mm (0.24 in) inside diameter drain lines, say, 0.5 m (19.7 in) in length, would contain $2m_3 \approx 0.024$ kg (0.053 lb) of oil. The dominance would be more notable if the oil drained through small-bore tubes because of the significant increase in the *effective* mass of the drain line oil. If the

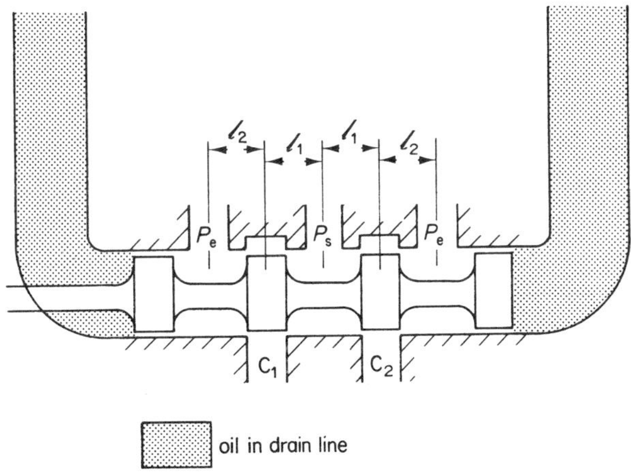


Figure A.4

line containing mass $m_3$ of fluid had a diameter $1/k$ times the spool diameter, then the effective mass would approximate to $k^4 m_3$ (see equation 3.5). For example, a 3 mm (0.12 in) internal diameter drain line 0.5 m (19.7 in) in length would contain about 0.003 kg (0.0066 lb) of oil but, associated with a 6 mm (0.24 in) diameter spool, its effective mass referred to the spool would equal about $2^4 \times 0.003$ or roughly 0.05 kg (0.11 lb). Hence the effective value of $2m_3$ would be 0.1 kg (0.22 lb) compared with a spool mass of, perhaps, 0.005 kg (0.011 lb).

Some types of spool valve can be made with generously sized upstream and downstream drain ports connected together in order to minimise the oil mass involved. Other types, however, particularly those for pilot operation (whether for metering or for on–off control) need small-bore connections or long lines or both.


### A.3  Frictional Forces

Some estimate of the frictional force of a spool sliding axially in its bore may be made using the basic concept of viscosity. With oil completely filling the annular gap of length $\ell_s$ between spool and bore and having constant absolute viscosity $\mu$, and taking both spool and bore to be axially aligned axially straight and of truly circular section, there are two main situations. The first is with the spool located in the exact centre of the bore with radial clearance $c_r$ constant over the whole periphery. The second is with the spool aligned axially but eccentric radially, that is with its axis displaced radially by distance $\varepsilon$ from the bore centre line.

In both cases, the surface area is $\pi d_1 \ell_s$ and, in the first case, the velocity gradient is taken as constant and equal to $x/c_r$. In the second case, velocity gradients vary round the periphery but are taken to be constant along any line. The results are

$$\text{for concentric spool } F_f = \mu \pi d_1 \ell_s \frac{\dot{x}}{c_r} \tag{A.5}$$

$$\text{for eccentric spool } F_f = \mu \pi d_1 \ell_s \frac{\dot{x}}{c_r \{1 - (\varepsilon/c_r)^2\}^{1/2}} \tag{A.6}$$

(Note that kinematic viscosity $v$ in stokes or centistokes (1 cSt $= 10^{-6}$ m²/s) multiplied by specific gravity gives numerically the absolute viscosity $\mu$ in poises or centipoises (1 cP $= 10^{-3}$ N s/m²) respectively and that, for British units, poises multiplied by 0.000 014 5 give $\mu$ in reynolds to use with in lb s units). Numerically for a spool of length 38 mm (1.496 in), of diameter 12 mm (0.472 in), with radial clearance 0.01 mm (0.000 39 in), which is moving at a steady velocity of 75 mm/s (2.95 in/s), if the oil viscosity is 50 cSt, and the oil specific gravity is 0.87, then the forces predicted are for the concentric spool

$$F_f = 43.5 \times 10^{-3} \times \pi \times 12 \times 10^{-3} \times 38 \times 10^{-3} \times \frac{75 \times 10^{-3}}{0.01 \times 10^{-3}} = 0.47 \text{ N}$$

or in British units

$$F_f = 0.87 \times 0.5 \times 0.000\,014\,5 \times \pi \times 0.472 \times 1.496 \times \frac{2.95}{0.000\,39} = 0.106 \text{ lbf}$$

and for an eccentricity of 0.008 mm

$$F_f = 0.47\frac{1}{\{1-(0.8)^2\}^{1/2}} = 1.31 \text{ N } (0.29 \text{ lbf})$$

or an eccentricity of 0.009 mm

$$F_f = 0.47\frac{1}{\{1-(0.9)^2\}^{1/2}} = 2.47 \text{ N } (0.56 \text{ lbf})$$

Such calculations tend to underestimate friction forces unless precautions are taken to prevent the spool jamming against the side of the bore. Such precautions concern hydraulic lock, straightness of axes, circularity and surface finish.


## A.4  Summary

By taking the pressure drop at each of the two active ports as $P_s/2$ and the resulting momentum forces as equal (and additive), then for the type of valve shown in figure A.4, but with a centring spring of rate $K$, the force acting on the spool is predicted as

$$(m_1 + m_2 + k^4 m_3)\ddot{x} + \left\{ (\ell_1 - \ell_2)C_d \pi d_1 (\rho P_s)^{1/2} + \frac{\mu \pi d_1 \ell_s}{c_r} \right\}\dot{x}$$

$$+ \left\{ \frac{7}{16}\pi d_1 P_s + K \right\} x \qquad (A.7)$$

(written $m'\ddot{x} + f'\dot{x} + k'x$ in equation 9.1). Fluid friction in the valve chamber is neglected (see horenz and Stringer (1966–7)).

The above analysis represents an attempt to distil the main factors affecting the motion of a valve spool in its bore. Unfortunately a number of things tend to upset such calculations. Examples of when this happens are as follows: if the spool is tilted slightly; if the spool or bore is irregular in surface finish; if oil flow patterns in the valve chamber change; if the angle at which jets enter or leave the valve chamber change.

**Problem**

1 The cylindrical spool of the simple valve illustrated in figure A.1 slides in its bore when an external axial force is applied. Estimate the magnitude of the force needed to hold the spool stationary if the valve has full periphery annular ports given that $d_1 = \frac{1}{2}$ in (12.7 mm); $x = 0.01$ in (0.254 mm); $P_a = 3000$ lb/in² (206.9 bar). Discuss the probable accuracy of your result due to, for example, neglecting friction and assuming a jet angle.

(21.1 lb or 93.8 N for $\theta = 69°$.)

# Appendix B Three-way Valve Systems

Methods similar to those used in chapter 7 can be employed to predict the flow through valves of the three-way (as opposed to four-way) type. Methods similar to those used in chapter 8 can be applied to the control of pistons by three-way valves.

## B.1   Valves

A similar equation to equation 7.1 (for four-way valves) is used to express the flowrate through three-way valves, namely

$$q = K_q x - K_c P_m'  \tag{B.1}$$

but in this case $P_m'$ is usually defined for convenience as $P_m' = P_c - P_s/2$ because this form is most useful when dealing with the half-area differential piston systems (see below) which are most commonly adopted with three-way valve control.

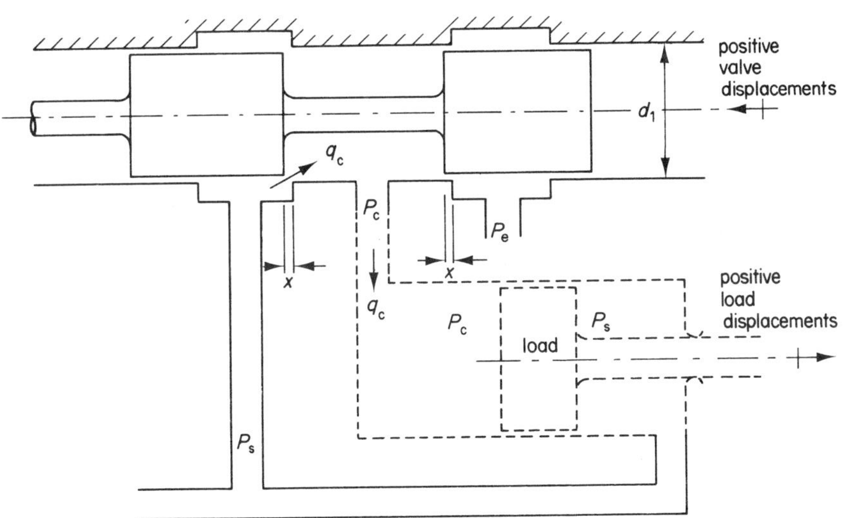


Figure B.1

### B.1.1  *Critical Centre Type*

The central spool position just closes both the supply and the exhaust port of the valve. A displacement of the spool to the left in figure B.1 causes fluid to be metered into the ram chamber from the supply, whereas one to the right causes metering of fluid from the chamber to exhaust. Referring to figure B.1, the oil flowrate $q_c$ is given by

$$q_c = C_d \pi d_1 x (P_s - P_c)^{1/2} \left(\frac{2}{\rho}\right)^{1/2} \quad \text{for } x + \text{ve}$$

or

$$q_c = C_d \pi d_1 x (P_c - P_e)^{1/2} \left(\frac{2}{\rho}\right)^{1/2} \quad \text{for } x - \text{ve}$$

and substituting $P_m' = P_c - P_s/2$

$$q_c = C_d \pi d_1 x \left(\frac{P_s}{2} - P_m'\right)^{1/2} \left(\frac{2}{\rho}\right)^{1/2} \quad \text{for } x + \text{ve}$$

or

$$q_c = C_d \pi d_1 x \left(\frac{P_s}{2} + P_m'\right)^{1/2} \left(\frac{2}{\rho}\right)^{1/2} \quad \text{for } x - \text{ve}$$

which may be approximated; noting that

$$\left(\frac{P_s}{2} \pm P_m'\right)^{1/2} = \left\{\frac{P_s}{2}\left(1 \pm \frac{2P_m'}{P_s}\right)\right\}^{1/2} \approx \left(\frac{P_s}{2}\right)^{1/2}\left(1 \pm \frac{1}{2}\frac{2P_m'}{P_s}\right)$$

as follows:

$$q_c = C_d \pi d_1 x (P_s)^{1/2}\left(\frac{1}{\rho}\right)^{1/2} \pm C_d \pi d_1 x (P_s)^{1/2}\frac{1}{P_s} P_m' \tag{B.2}$$

(where the $+$ ve sign is associated with negative values of $x$ and the $-$ ve sign with positive ones). Hence (using $C_d(1/\rho)^{1/2} = 6.7$ for SI units with pressures in bars)

$$K_q = 6.7\pi d_1 (P_r)^{1/2} \tag{B.3}$$

$$K_c = 6.7\pi d_1 |x|\frac{P_s}{P_s} \tag{B.4}$$

three-way closed centre in SI units. (Note that $K_p = P_s/x$ and the value of $K_c$ would be predicted to be zero at null.)

### B.1.2  *Open Centre Type (Underlapped Three-way Valve)*

Referring to figure B.2

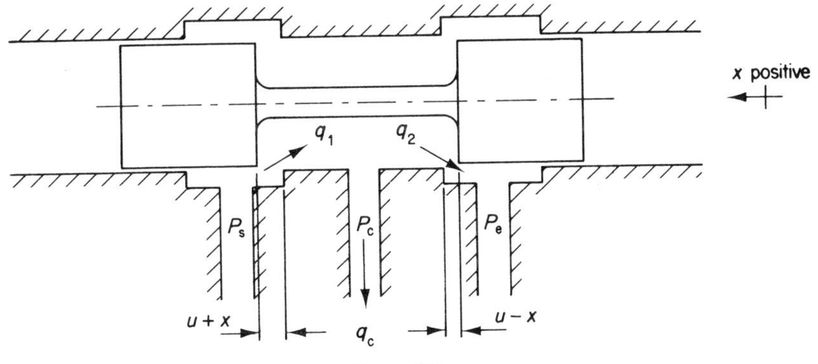


Figure B.2

$$q_1 = C_{\mathrm{d}}\pi d_1(u + x)(P_{\mathrm{s}} - P_{\mathrm{c}})^{1/2}\left(\frac{2}{\rho}\right)^{1/2}$$

$$\approx C_{\mathrm{d}}\pi d_1(u + x)(P_{\mathrm{s}})^{1/2}\left(\frac{1}{\rho}\right)^{1/2}\left(1 - \frac{P_{\mathrm{m}}}{P_{\mathrm{s}}}\right)$$

$$q_2 = C_{\mathrm{d}}\pi d_1(u - x)(P_{\mathrm{c}})^{1/2}\left(\frac{2}{\rho}\right)^{1/2}$$

$$\approx C_{\mathrm{d}}\pi d_1(u - x)(P_{\mathrm{s}})^{1/2}\left(\frac{1}{\rho}\right)^{1/2}\left(1 + \frac{P_{\mathrm{m}}}{P_{\mathrm{s}}}\right)$$

$$q_{\mathrm{c}} = q_1 - q_2$$

$$\approx C_{\mathrm{d}}\pi d_1(P_{\mathrm{s}})^{1/2}\left(\frac{1}{\rho}\right)^{1/2}2x - C_{\mathrm{d}}\pi d_1(P_{\mathrm{s}})^{1/2}\left(\frac{1}{\rho}\right)^{1/2}\frac{2u}{P_{\mathrm{s}}}P_{\mathrm{m}}$$

and

$$K_q = 13.4\pi d_1(P_{\mathrm{s}})^{1/2} \tag{B.5}$$

$$K_{\mathrm{c}} = 13.4\pi d_1 u\frac{(P_{\mathrm{s}})^{1/2}}{P_{\mathrm{s}}} \tag{B.6}$$

three-way open centre. Note also $K_{\mathrm{p}} = P_{\mathrm{s}}/u$.

## B.2 Three-way Valve System

The three-way valve system shown in figure B.3 is similar to the four-way system considered previously (see figure 8.1).

This system relies on the differential ram area for its operation. Fluid is metered into and out of the larger-area chamber. At the balance point,

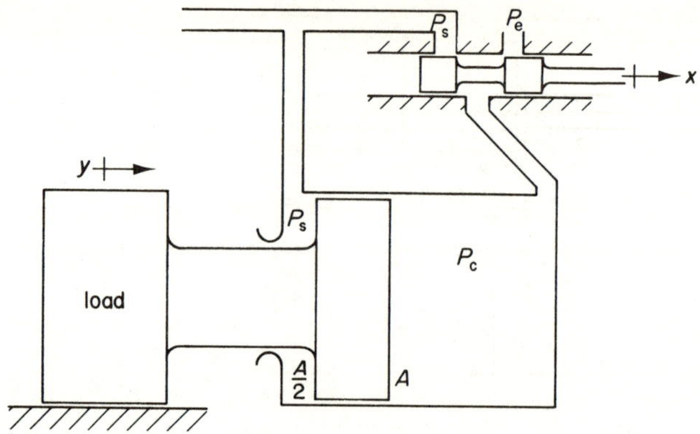


Figure B.3  Piston with three-way valve

the valve would be central in the diagram and the steady state control pressure would equal one-half the supply pressure. A movement of the spool to the left would cause a rise in the control pressure and a metering of fluid into the larger chamber. With the spool to the right (as illustrated in figure B.3) fluid would be being metered from the larger chamber and the control pressure $P_c$ would be lowered. The piston could exert the same maximum force to the left (when $P_c = P_s$) as to the right (when $P_c = 0$). This force is $\frac{1}{2} A P_s$.

Motions of the load caused by motions of the spool will be considered, assuming for simplicity that the load consists purely of a mass $M$. It will also be assumed that there is no leakage.

The flowrate to or from the valve may be written

$$q = K_q x - K_c P_m'  \tag{B.1}$$

(where $P_m' = P_c - P_s/2$ for this type of system).

This flowrate causes motion and also compression of the fluid

$$q = A \frac{\mathrm{d}y}{\mathrm{d}t} + \frac{V}{\beta} \frac{\mathrm{d}P_c}{\mathrm{d}t}  \tag{B.7}$$

where it is advisable to take $V$ as $V_t$, the maximum possible volume of the larger chamber (with the piston over to the left in the diagram) because this gives the lowest stiffness to the oil 'spring'.

The balance of forces, for a purely inertia load, is

$$A\left(P_c - \frac{P_s}{2}\right) = A P_m' = M \frac{\mathrm{d}^2 y}{\mathrm{d}t^2}  \tag{B.8}$$

and, differentiating with respect to time,

$$A \frac{\mathrm{d}(P_\mathrm{c} - P_\mathrm{s}/2)}{\mathrm{d}t} = A \frac{\mathrm{d}P_\mathrm{c}}{\mathrm{d}t} = M \frac{\mathrm{d}^3 y}{\mathrm{d}t^3} \qquad \text{(B.9)}$$

Equating B.1 with B.7 after substituting from B.8 in the second term of B.1 and from B.9 in the second term of B.7 and dividing by $A$ gives

$$\frac{K_q x}{A} = \frac{V_\mathrm{t}}{\beta} \frac{M}{A^2} \frac{\mathrm{d}^3 y}{\mathrm{d}t^3} + K_\mathrm{c} \frac{M}{A^2} \frac{\mathrm{d}^2 y}{\mathrm{d}t^2} + \frac{\mathrm{d}y}{\mathrm{d}t} \qquad \text{(B.10)}$$

(for a three-way valve system with pure inertia load) which is a similar equation to that of a comparable four-way valve system (equation 8.4) except for the first term on the right.

The operational relation between piston speed and valve displacement is the same as that for the systems considered before, but the hydraulic frequency is lower (for given $A$, $V_\mathrm{t}$ and $M$) mainly because there is only one volume of oil trapped between piston and valve (with four-way valve systems, there are two volumes and their stiffnesses are additive). The relation is

$$\frac{Dy}{x} = \frac{K_q/A}{(1/\omega_\mathrm{h}^2)D^2 + (2\zeta/\omega_\mathrm{h})D + 1} \qquad \text{(B.11)}$$

(same as equation 8.5) where

$$\omega_\mathrm{h}^2 = \frac{\beta}{V_\mathrm{t}} \frac{A^2}{M}$$

(for three-way valve system) and

$$\frac{2\zeta}{\omega_\mathrm{h}} = K_\mathrm{c} \frac{M}{A^2}$$

(with no leakage, friction or other 'damping').

## B.3   Three-way Valve Servo

A three-way valve position control system is shown in figure B.4.

A displacement of the input $\theta_\mathrm{i}$ to the right in figure B.4 would connect the larger area to exhaust ($P_\mathrm{e} = 0$), thereby causing the cylinder and load to move towards the right and to bring the valve to its closed position.

### B.3.1   *Governing Equation*

A differential equation is required to represent the system. A method of analysis similar to that used for the three-way system (figure B.3) is relevant. In fact all the equations developed for the three-way valve-controlled ram system are directly applicable to this system and need not be repeated.

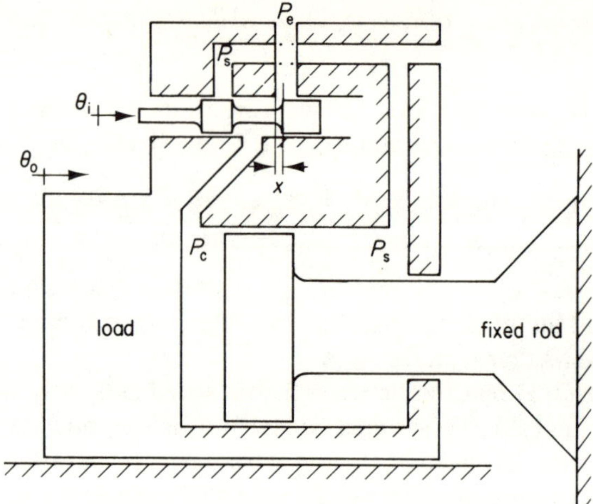



Figure B.4 Position control system with three-way valve

The flowrate through the valve is as shown in equation B.1; the motion is of the cylinder rather than the ram but otherwise equation B.7 also applies.

Assuming a pure inertia load, equations B.8 and B.9 apply directly noting that $M$ is the *total* moving mass including the cylinder and valve body in this case. The individual equations apply direct, so the combined equation B.10 must also apply. In equation B.10 the symbol $y$ is replaced by $\theta_o$.

The basic difference between the system sketched in figure B.4 and that in figure B.3 is that the displacements of the load and the valve spool are no longer independent. For the system of figure B.4, these are related by the equation $x = \theta_i - \theta_o$.

By substituting $x = \theta_i - \theta_o$ in equation B.10 we obtain the governing equation of the system shown in figure B.4 which is

$$\frac{K_q}{A}\theta_i = \frac{V_t}{\beta}\frac{M}{A^2}\frac{d^3\theta_o}{dt^3} + K_c\frac{M}{A^2}\frac{d^2\theta_o}{dt^2} + \frac{d\theta_o}{dt} + \frac{K_q}{A}\theta_o \tag{B.12}$$

(for the three-way valve system of figure B.4).

This governing equation is of the form

$$(a_0 D^3 + a_1 D^2 + a_2 D + a_3)\theta_o = a_3\theta_i \tag{B.13}$$

(same as equation 8.8) where

$$a_1 = \frac{2\zeta}{\omega_h} = K_c\frac{M}{A^2}, \qquad\qquad a_3 = \frac{K_q}{A}$$

$$a_0 = \frac{1}{\omega_h^2} = \frac{\beta A^2}{V_t M} \text{ (three-way system)}, \ a_2 = 1$$

The block diagram for this three-way valve system (figure B.4) would be the same as that for the four-way valve system and is given as figure 8.3. However, the numerical values of the coefficients ($\omega_h$, $\zeta$, $K_q$, $A$) would be different.

# Appendix C   Special Purpose Valves

The analyses given below are primarily to illustrate the type of linearised approach which might prove useful for special purpose hydraulic valves (see also Foster (1961) and Takenaka and Vrata (1968)).

The dynamic characteristics of hydraulic devices such as relief valves or flow control valves are not very well understood. For design purposes it would be useful to know the likely influence of spool mass, spring rate, orifice size or other parameter on the response of any particular valve to changes in pressure or flowrate or to other disturbances. A pressure control valve and a flow control valve are considered below.

### C.1   Poppet Valves

Each of the two devices to be considered employs a poppet valve (see figure C.1) and their characteristics will be influenced by the flow patterns existing at or near the valve seat. For the purposes of analysis we shall assume that flow through the valve opening takes the form of a jet (of annular section) which follows the face of the poppet. Actual flow patterns differ from this (see McCloy and McGuigan (1965)).

Referring to figure C.1, for poppet displacement $x$, the area of flow is

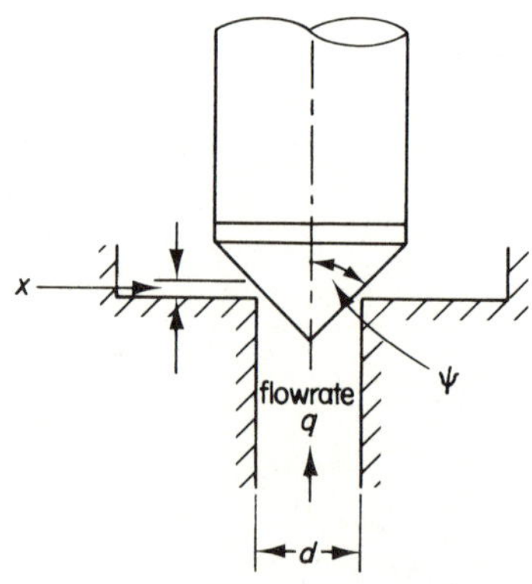



Figure C.1  Poppet valve

$x\pi d \sin \psi$ and the area of the jet $C_d x\pi d \sin \psi$, where $C_d$ is a flow coefficient. For a pressure drop of $\Delta P$ across the orifice formed between seat and poppet, the fluid velocity $v$ is taken as equal to

$$\left(\frac{2\,\Delta P}{\rho}\right)^{1/2}$$

(see chapter 1) and the volume flowrate as

$$q = C_d x\pi d \sin \psi \left(\frac{2\,\Delta P}{\rho}\right)^{1/2} \tag{C.1}$$

The momentum of the jet will have an axial component equal to $\dot{m}v \cos \psi$ (where $\dot{m}$ is the mass flowrate) which may be written

$$\rho C_d x\pi d \sin \psi \left(\frac{2\,\Delta P}{\rho}\right)^{1/2} \left(\frac{2\,\Delta P}{\rho}\right)^{1/2} \cos \psi$$

or

$$\text{axial component of jet momentum} = \lambda \Delta P x \tag{C.2}$$

where

$$\lambda = C_d \pi d \sin 2\psi$$

## C.2 Single-stage Relief Valve

A spring-loaded valve is illustrated in figure C.2.

The pressure applied to the valve is designated $P$, and it is the effects of small changes in this pressure from $P_0$ to $P_0 + \overset{*}{P}$ which are to be considered. The outlet (exhaust) pressure is assumed to be zero throughout.

With the poppet closed (that is $x = 0$) there will be some spring preload force holding it down which will be designated $F$.

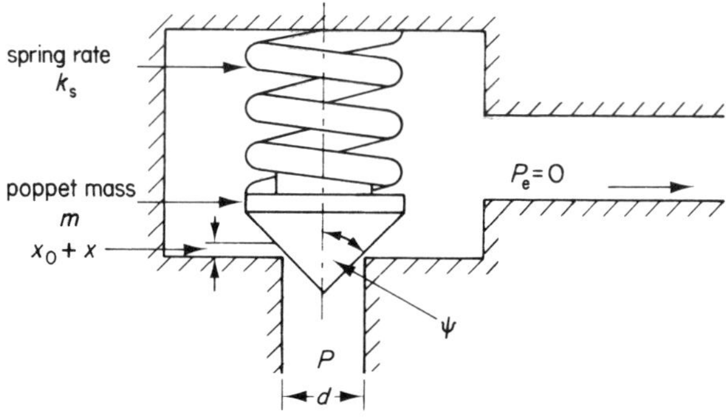


Figure C.2 Relief valve

Under steady state conditions with a greater pressure applied than that needed to overcome the preload, and with the valve partly open and the poppet stationary, there will be a steady state balance of forces relating this pressure to the poppet displacement. For applied pressure $P_0$ and poppet displacement $x_0$, the relation is

$$P_0 \frac{\pi d^2}{4} = F + k_s x_0 + \lambda P_0 x_0 \qquad (C.3)$$

(if the approach velocity is negligible—note that $\lambda$ is given in equation C.2 and that fluid frictional drag across the poppet face is neglected). For some other steady state position of the poppet (displacement $x_0 + \overset{*}{x}$), another pressure $P_0 + \overset{*}{P}$ would occur according to

$$(P_0 + \overset{*}{P}) \frac{\pi d^2}{4} = F + k_s(x_0 + \overset{*}{x}) + \lambda(P_0 + \overset{*}{P})(x_0 + \overset{*}{x}) \qquad (C.3a)$$

(equations C.3 and C.3a refer to steady state with poppet stationary).

Under dynamic conditions with the poppet moving, the balance of forces has to take into account the effective mass of the poppet $m$ (which would include one-third the mass of the spring) and any damping (assumed viscous of rate $f$) as well as the spring, pressure and momentum forces. For a pressure $P_0 + \overset{*}{P}$ and a poppet displacement $x_0 + x$ (where $x \neq \overset{*}{x}$ and both $\overset{*}{P}$ and $x$ are assumed to be small) the balance of forces becomes

$$(P_0 + \overset{*}{P}) \frac{\pi d^2}{4} = F + k_s x_0 + \lambda P_0 x_0 + \lambda x_0 \overset{*}{P} + (k_s + \lambda P_0)x + fDx$$

$$+ mD^2 x \quad (C.4)$$

(assuming that terms involving $\overset{*}{P}x$ are negligibly small). Subtracting equation C.3 from C.4 gives

$$\overset{*}{P} \frac{\pi d^2}{4} - \overset{*}{P} \lambda x_0 = (k_s + \lambda P_0)x + fDx + mD^2 x$$

indicating a second-order relation between changes in pressure $\overset{*}{P}$ and changes in poppet displacement $x$ given by

$$\frac{x}{\overset{*}{P}} = \frac{K}{(1/\omega_n^2)D^2 + (2\zeta/\omega_n)D + 1} \qquad (C.5)$$

where

$$K = \frac{\pi d^2/4 - \lambda x_0}{k_s + \lambda P_0}$$

and

$$\omega_n^2 = \frac{k_s + \lambda P_0}{m}$$

This relation suggests that any oscillation of the poppet will be associated with a much stiffer 'spring' than the physical spring constant would suggest. As a numerical example, consider a valve of diameter $d = 6$ mm (0.24 in) for use at a nominal pressure of $P_0 = 70$ bar (1015 lb/in$^2$). The projected area $\pi d^2/4$ equals about 28 mm$^2$ (0.04 in$^2$), so the value of $F$ is approximately 20 N (45 lb) and, if this is obtained by initially compressing the spring 10 mm (0.39 in), the spring rate would have a value $k_s \approx 20$ N/mm (114 lb/in). Assume the poppet has a 90° cone angle ($\psi = 45°$) and the flow coefficient $C_d = 0.7$, then $\lambda$ for the valve is about 13.2 mm (0.52 in) and $\lambda P_0 = 0.924 \times 10^5$ N/m or 92.4 N/mm (527 lb/in). The effective spring rate would be not 20 but 112.4 N/mm (641 lb/in). If the effective poppet mass were 0.01 kg (0.022 lb), then the natural frequency $\omega_n$ of poppet oscillations would be 533 Hz.

## C.3  A Flow Control Valve

Figure C.3 illustrates a valve which is designed to pass a constant flowrate of fluid despite fluctuations of the inlet and the outlet pressures. This type of device is often called a 'pressure-compensated flow control valve'.

The device has two orifices in series. One is preset manually to select the desired flowrate. The other (in this case annular) orifice varies with the pressure difference across the valve—a decrease in the pressure difference causes an increase in the control orifice size. The aim is to keep the flowrate constant by maintaining a constant pressure difference across the preset orifice.

For analysis, changes in the outlet pressure $P_L$ will represent the disturbances externally imposed on the device with the inlet (supply) pressure

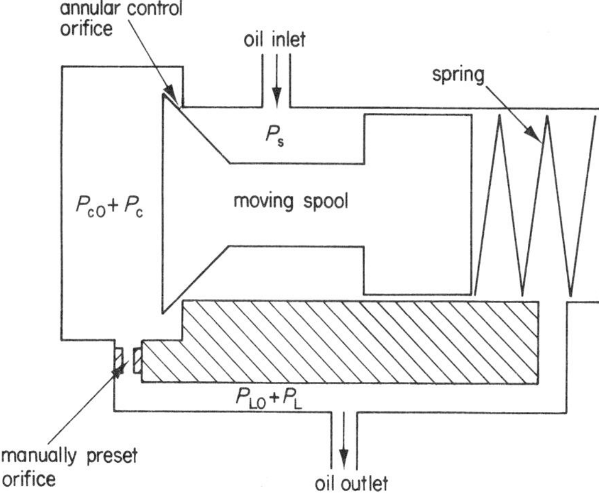


Figure C.3  Pressure-compensated flow control valve

$P_s$ assumed constant. This would simulate a 'meter in' control with the device used on a system. Normal operation only occurs with the poppet valve partly open and this analysis concerns small changes in this valve opening. The fully open or fully closed conditions are not dealt with. The datum for the spring force acting on the spool is taken from some arbitrary position of the spool represented by the poppet (that is control orifice) opening of $x_0$.

### C.3.1   Forces

The steady state balance of forces for some equilibrium operating position with the poppet stationary and open distance $x_0$, for supply pressure $P_s$, outlet pressure $P_{L0}$ and consequent chamber pressure $P_{c0}$ is given by

$$(P_{c0} - P_{L0})A_p = F - \lambda(P_s - P_{c0})x_0 \qquad (C.6)$$

(assuming approach velocity of fluid entering control orifice is negligible and noting that $F$ is the spring force for opening $x_0$ and also that $\lambda$ is given by equation C.2).

Under dynamic conditions with the spool in motion for outlet pressure $P_{L0} + P_L$, with the (instantaneous) spool position distance $x$ to the *left* of its initial position, noting that $P_{c0} + P_c$ is the (instantaneous) chamber pressure, the balance of forces is given by

$$\{(P_{c0} + P_c) - (P_{L0} + P_L)\}A_p =$$
$$F - k_s x - \lambda\{P_s - (P_{c0} + P_c)\}(x_0 + x) - fDx - mD^2x \qquad (C.7)$$

where $m$ is the effective moving mass and $f$ the effective viscous damping.

Assuming that only small changes of variable are being dealt with so that $P_c$, $P_L$ and $x$ are small and terms involving $P_c x$ may be neglected, then subtracting equation C.7 and C.6 gives

$$-P_c = \frac{k_s + \lambda(P_s - P_{c0})}{A_p - \lambda x_0}x + \frac{f}{A_p - \lambda x_0}Dx + \frac{m}{A_p - \lambda x_0}D^2x$$

$$-\frac{A_p}{A_p - \lambda x_0}P_L \qquad (C.8)$$

### C.3.2   Flowrates

Under steady state conditions, the flowrate through the control orifice must equal that through the preset orifice. Hence for the steady conditions previously used with inlet pressure constant at $P_s$, outlet pressure $P_{L0}$ chamber pressure $P_{c0}$ and control orifice opening $x_0$, using suffix e for the preset orifice we have

$$q_0 = C_d x_0 \pi d \left\{ \frac{2(P_s - P_{c0})}{\rho} \right\}^{1/2} = C_{de} a_e \left\{ \frac{2(P_{co} - P_{L0})}{\rho} \right\}^{1/2} \quad \text{(C.9)}$$

Under dynamic conditions, the two flowrates will no longer be equal—the flowrate through the control orifice will be designated $q_0 + q_i$ and that through the preset orifice $q_0 + q_e$. The difference between the two flowrates (into and out from the valve chamber) causes compression of the oil in the valve chamber or

$$q_i - q_e = \frac{V}{\beta} DP_c \quad \text{(C.10)}$$

(equation C.10 is based on the assumption that $V$ may be regarded as constant).

Under the dynamic conditions previously used (see equation C.7), the flowrate through the control orifice may be written

$$q_0 + q_i = C_d (x_0 + x) \pi d \left( \frac{2}{\rho} \right)^{1/2} \{ P_s - (P_{c0} + P_c) \}^{1/2} + A_p Dx \quad \text{(C.11)}$$

and subtracting equation C.9 from C.11, neglecting terms involving $P_c x$ as being negligibly small,

$$q_i = -\frac{q_0}{2} \frac{P_c}{P_s - P_{c0}} + \frac{q_0}{x_0} x + A_p Dx \quad \text{(C.12)}$$

and similarly for the change in flowrate through the preset orifice

$$q_e = \frac{q_0}{2} \frac{P_c}{P_{c0} - P_{L0}} - \frac{q_0}{2} \frac{P_L}{P_{c0} - P_{L0}} \quad \text{(C.13)}$$

Subtracting equation C.13 from C.12 gives

$$q_i - q_e = A_p Dx + \frac{q_0}{x_0} x + \frac{q_0}{2} \frac{P_L}{P_{c0} - P_{L0}}$$

$$- \frac{q_0}{2} \left( \frac{1}{P_s - P_{c0}} + \frac{1}{P_{c0} - P_{L0}} \right) P_c \quad \text{(C.14)}$$

Equating C.10 and C.14 after substituting equation C.8 for $P_c$ and, after differentiating, for $DP_c$ to obtain $P_c$ and $DP_c$ in terms of $P_L$ and $x$ leads to the (closed loop) relation between displacements of the spool and changes of the outlet pressure

$$a_0 D^3 x + a_1 D^2 x + a_2 Dx + a_3 x = KP_L + TDP_L \quad \text{(C.15)}$$

or

$$\frac{x}{P_L} = \frac{K + TD}{a_0 D^3 + a_1 D^2 + a_2 D + a_3}$$

where

$$a_3 = k'_s Q' + \frac{q_0}{x_0}(A_p - \lambda x_0)$$

$$a_2 = f Q' + k'_s \frac{V}{\beta} + A_p(A_p - \lambda x_0)$$

$$a_1 = mQ' + f\frac{V}{\beta}$$

$$a_0 = m\frac{V}{\beta}$$

$$T = A_p\frac{V}{\beta}$$

$$K = A_p\frac{q_0}{2(P_s - P_{co})} + \lambda x_0\frac{q_0}{2(P_{co} - P_{LO})}$$

using

$$Q' = \frac{q_0}{2(P_s - P_{co})} + \frac{q_0}{2(P_{co} - P_{LO})}$$

$$k'_s = k_s + \lambda(P_s - P_{co})$$

This linear relation is obviously difficult to apply particularly as the coefficients depend on the initial conditions (for example on the nominal flowrate $q_0$) which might be classed as the 'normal' operating conditions. In practice valves of this type can be unsatisfactory at initial starting (when $x_0$ may be large). The relation does indicate that a valve will be stable if $a_1 a_2$ is greater than $a_3 a_0$ and confirms that increasing $f$, the viscous damping, will help to stabilise the valve, increasing both $a_1$ and $a_2$.

# Appendix D  Numerical Examples

Examples are given below to illustrate a selection of the calculation methods described in the text. These are in the form of possible answers to two questions.

## D.1  Question (i)

A double-acting power cylinder of 300 mm (11.81 in) stroke and 75 mm (2.95 in) internal diameter $d_2$ has a wall thickness $t$ of 6 mm (0.24 in). Young's modulus $E$ for the cylinder material is $9 \times 10^{10}$ N/m$^2$ ($13 \times 10^6$ lb/in$^2$) and Poisson's ratio $v$ is $\frac{1}{3}$. The 25 mm (0.98 in) long piston is fitted in the cylinder without seals and with an initial radial clearance of 0.025 mm (0.001 in), that is under no pressure.

The oil in the cylinder has bulk modulus $\beta_o$ of $14 \times 10^8$ N/m$^2$ (203 050 lb/in$^2$) viscosity 25 cSt and specific gravity 0.8. It contains the equivalent of 1% non-dissolved air.

The double piston rod has a diameter $d_3$ of 30 mm (1.18 in) and the total moving mass $M$ (including piston rod and rigidly attached load) is 1000 kg (2538 lb).

The piston is so controlled that 175 bar (pressure $P$) is maintained on either side of the piston under equilibrium conditions.

Estimate the hydraulic frequency of the actuator and its load. Estimate also the damping factor due to leakage past the piston; and the damping factor due to leakage plus viscous drag.

(Note dynamic viscosity $\mu = 25 \times 0.8$ cP.)

### D.1.1  *Calculation of Answers to Question (i)*

$$\frac{V_t M}{4\beta_e A^2} = \frac{1}{\omega_h^2} \quad ; \quad L\frac{M}{A^2} = \frac{2\zeta}{\omega_h}$$

or

$$\left(L + \frac{fV_t}{4\beta_e M}\right)\frac{M}{A^2} = \frac{2\zeta'}{\omega_h}$$

For the cylinder distention

$$\frac{1}{\beta_c} = \frac{2}{E}\left(\frac{d_1^2 + d_2^2}{d_1^2 - d_2^2} + v\right)$$

$$= \frac{2}{9 \times 10^{10}}\left(\frac{87^2 + 75^2}{87^2 - 75^2} + \frac{1}{3}\right)$$

Hence

$$\beta_c = 63.2 \times 10^8 \text{ N/m}^2$$

or in British units

$$\frac{2}{13.05 \times 10^6}\left(\frac{3.43^2 + 2.95^2}{3.43^2 - 2.95^2} + \frac{1}{3}\right)$$

giving

$$\beta_c = 0.93 \times 10^6 \text{ lb/in}^2$$

By 'thin-wall' calculation

$$\frac{1}{\beta_c} = \frac{1}{E}\frac{d_1 + d_2}{d_1 - d_2} = \frac{1}{9 \times 10^{10}}\frac{87 + 75}{87 - 75}$$

Hence

$$\beta_c = 66.7 \times 10^8 \text{ N/m}^2$$

or in British units

$$\frac{1}{13.05 \times 10^6}\frac{3.43 + 2.95}{3.43 - 2.95}$$

giving

$$\beta_c = 0.98 \times 10^6 \text{ lb/in}^2$$

For the air content

$$\frac{1}{\beta_a} \text{ approximates to } \frac{V_{air}}{V_{oil}}\frac{1}{P}$$

that is

$$\frac{1}{100}\frac{1}{175 \times 10^5}$$

so

$$\beta_a \approx 17.5 \times 10^8 \text{ N/m}^2 \ (253\,800 \text{ lb/in}^2)$$

Hence the effective bulk modulus $\beta_e$

$$\frac{1}{\beta_e} = \frac{1}{14 \times 10^8} + \frac{1}{63.2 \times 10^8} + \frac{1}{17.5 \times 10^8}$$

so that

$$\beta_e = 6.9 \times 10^8 \text{ N/m}^2$$

or in British units

$$\frac{1}{203\,050} + \frac{1}{0.93 \times 10^6} + \frac{1}{253\,800}$$

so

$$\beta_e = 100\,600 \text{ lb/in}^2$$

The minimum stiffness with the actuator in midstroke

$$\frac{4A\beta_e}{l} = \frac{4\{(\pi/4)75^2 - (\pi/4)30^2\}\,10^{-6} \times 6.9 \times 10^8}{300 \times 10^{-3}}$$

$$= 0.341 \times 10^8 \text{ N/m}$$

or in British units

$$\frac{4\{(\pi/4)(2.95)^2 - (\pi/4)(1.18)^2\}\,100\,600}{11.81} = 195\,600 \text{ lb/in}$$

Hence

$$\omega_h = \left(\frac{0.341 \times 10^8}{1000}\right)^{1/2} = 184.7 \text{ rad/s or } 29.4 \text{ Hz}$$

or, noting that the mass for in lb s units is 2204.6/386,

$$\omega_h = \left(\frac{195\,600}{5.71}\right)^{1/2} = 185 \text{ rad/s}$$

Check using equation 8.5 noting that area $A = 3.71 \times 10^{-3}$ m$^2$ and volume $V_t \approx$ area $\times$ stroke or $1.11 \times 10^{-3}$ m$^3$

$$\omega_h = \left\{\frac{4 \times 6.9 \times 10^8 \times (3.71 \times 10^{-3})^2}{1.11 \times 10^{-3} \times 1000}\right\}^{1/2} = 185 \text{ rad/s}$$

or, noting $A = 5.75$ in$^2$ and $V_t \approx 67.9$ in$^3$,

$$\omega_h = \left(\frac{4 \times 100\,600 \times 33.1 \times 386}{67.9 \times 2204.6}\right)^{1/2} = 185 \text{ rad/s}$$

*Diameter* changes taken as half-rate of area changes, namely $63.2 \times 10^8$ N/m$^2$ ($0.93 \times 10^6$ lb/in$^2$), so diameter change for 175 bar (2538 lb/in$^2$) is

$$\frac{175 \times 10^5}{2 \times 63.2 \times 10^8} \times 75 \times 10^{-3} \approx 0.1 \times 10^{-3} \text{ m}$$

or in British units

$$\frac{2538 \times 2.95}{2 \times 0.93 \times 10^6} \approx 0.004 \text{ in}$$

so *radial* clearance increases from 0.025 mm to 0.075 mm (0.001 in to 0.003 in).

The calculation of $L$ from the formula for annulus

$$q = \frac{\Delta P \, \pi d c_r^{\,3}}{\ell 12 \mu} \text{ and } L = \frac{q}{\Delta P}$$

$$L = \frac{\pi \times 75 \times 10^{-3}(0.075 \times 10^{-3})^3}{25 \times 10^{-3} \times 12 \times 20 \times 10^{-3}}$$

$$= 1.66 \times 10^{-11} (\text{m}^3/\text{s})/(\text{N/m}^2)$$

$$= 0.001\,66 \text{ (l/s)/bar}$$

or in British units

$$L = \frac{\pi \times 2.95 \times (3 \times 10^{-3})^3}{1 \times 12 \times 20 \times 14.5 \times 10^{-8}} = 0.0072 \text{ (in}^3/\text{s)/(lb/in}^2)$$

Damping with $L$ alone

$$\frac{2\zeta}{\omega_h} = \frac{LM}{A^2}$$

$$\zeta = \frac{185 \times 1.66 \times 10^{-11} \times 1000}{2(3.71 \times 10^{-3})^2} \approx 0.1$$

or in British units

$$\zeta = \frac{185 \times 0.0072 \times 2538}{2 \times (5.75)^2 \times 386} \approx 0.1$$

Estimate viscous friction coefficient $f$

$$f = \frac{\mu \pi d \ell}{c_r} = \frac{20 \times 10^{-3} \times \pi \times 75 \times 10^{-3} \times 25 \times 10^{-3}}{0.075 \times 10^{-3}} = 1.6 \text{ N/(m/s)}$$

or in British units

$$\frac{20 \times 14.5 \times 10^{-8} \times \pi \times 2.95 \times 0.98}{0.003} = 0.009 \text{ lb/(in/s)}$$

a negligibly small value but to check the order of magnitude, compare the term $L$ with the term $fV/4M$, that is

$$\frac{1.6 \times 1.11 \times 10^{-3}}{4 \times 6.9 \times 10^8 \times 1000} = 6.44 \times 10^{-16} \text{ (whereas } L = 1.66 \times 10^{-11})$$

or in British units

$$\frac{0.009 \times 67.9 \times 386}{4 \times 100\,600 \times 2204} = 0.000\,000\,27 \quad (\text{whereas } L = 0.0072)$$

which demonstrate that the effect on the damping of the viscous friction between the (centralised) piston and its bore is negligible—no account has been taken of rod seal friction, however (see Healey and Stringer, 1968–9).


### D.1.2  *Answers to Question (i)*

$$\omega_h \approx 29\tfrac{1}{2} \text{ Hz}, \qquad \zeta \approx 0.1$$


## D.2  Question (ii)

An electrohydraulic servo valve of the flow control type is used to meter fluid to and from a hydraulic motor which controls the rectilinear position of a table. A lead screw driven direct by the motor shaft engages with a nut attached to the table in order to convert rotary into linear motion.

The table position is sensed by a potentiometer and the current to the servo valve coils is directly proportional to the difference between an input voltage (representing desired table position) and the voltage from the potentiometer.

Suggest a suitable supply voltage (in V/mm of table travel) for the potentiometer if the following data apply: amplifier gain $K_d = 6$ (that is 1 V error voltage causes 6 V to be applied on valve coils); valve coil impedance $= 20\ \Omega$ (assumed purely resistive); valve 'rating' based on one-third actual supply pressure is 0.008 l/s (0.49 in$^3$/s) for every milliampere of coil current; motor capacity $\delta_m = 0.005$ l/rad (0.305 in$^3$/rad); trapped oil volume between valve and motor $V_T = 0.15$ l (9.15 in$^3$); mass of table and parts moving with it 3500 kg (7716 lb); lead of screw $\mathscr{G} = 2.5$ mm/rad (0.098 in/rad).

Assume that some damping exists due, for example, to internal leakage and viscous friction. Take the *effective* bulk modulus $\beta$ for the oil as $11 \times 10^8$ N/m$^2$ (159 540 lb/in$^2$).


### D.2.1  *Calculation of Answer to Question (ii)*

$$K_I K_q = (3)^{1/2} \times 10^3 \times 0.008 \times 10^{-3} = 13.86 \times 10^{-3} (\text{m}^3/\text{s})/\text{A}$$

or in British units

$$(3)^{1/2} \times 10^3 \times 0.49 = 849 \ (\text{in}^3/\text{s})/\text{A}$$

$$J = MG^2 = 3500(2.5 \times 10^{-3})^2 = 0.022$$

or in British units $(7716/386)(0.098)^2 = 0.192$.

$$\omega_h = \left(\frac{4\beta\delta_m^2}{V_T J}\right)^{1/2}$$

$$= \left\{\frac{4 \times 11 \times 10^8 \times (0.005 \times 10^{-3})^2}{0.15 \times 10^{-3} \times 0.022}\right\}^{1/2}$$

$$= 183 \text{ rad/s} \approx 29 \text{ Hz}$$

or in British units

$$\left\{\frac{4 \times 159\,540 \times (0.305)^2}{9.15 \times 0.192}\right\}^{1/2} = 184 \text{ rad/s}$$

Hence $\omega_h$ is approximately 180 rad/s and a suitable value for the gain (that is $K_R K_e K_d K_I K_q / \delta_m$) would be, say, $\omega_h/5$ or $36 \text{ s}^{-1}$, that is

$$36 = \frac{1}{20} \times K_e \times 6 \times 13.86 \times 10^{-3} \times \frac{1}{0.005 \times 10^{-3}}$$

Therefore

$$K_e = 0.043 \text{ V/rad}$$

$$= 0.0172 \text{ V/mm} \text{ (or 17.2 V/m)}$$

or in British units

$$36 = \frac{K_e \times 6 \times 849}{20 \times 0.305}$$

therefore

$$K_e = 0.043 \text{ V/rad} \text{ or } 0.44 \text{ V/in (that is 17.3 V/m)}$$

### D.2.2   *Answer to Question (ii)*

About 1/60 V/mm of table travel.

# Appendix E   Hydraulic Lock

'Hydraulic lock' is the name given to the effect observed with some spool valves of the spool sticking in its bore when pressure is applied to the valve. Before applying pressure, the spool is found to move freely and the effect is not caused by particles lodged in the annular gap between the spool and the bore. The effect is associated with the distributions of pressure within this annular gap and can be explained qualitatively in terms of the effects of manufacturing imperfections on these pressure distributions. More quantitative information may be found in Manhajm and Sweeney (1955) and Dransfield (1967–8).

We make a general assumption that all types of imperfection due to manufacturing can be simply represented as a tapering of the spool relative to the bore. The nominal radial clearances between valve spools and their bores usually lie in the range 0.005 mm (0.0002 in) to 0.025 mm (0.001 in) and changes in diameter need only be of the order of, say, 0.001 mm (0.000 04 in) for the (conical) tapering effect to be significant. A plain cylindrical bore will now be considered ignoring any valve ports. Simplifying assumptions are also made about the flow of fluid. Laminar flow is assumed to occur along the annulus parallel to the cylindrical bore with zero flow in the peripheral direction. The spool and bore are regarded as having constant roughness over the whole of their surfaces and the temperature of the fluid is assumed constant throughout.

If the bore and the spool are perfectly straight and circular in section and if the spool is exactly central and parallel in the bore, then there is a constant radial clearance throughout the annulus. The rate at which the pressure decreases along the length of the annulus is constant and there is a straight line relation between pressure and distance along the length of the spool as shown in figure E.1. Even if the spool is not centrally placed, so long as it is parallel in the bore, the pressure distribution is still the same for any line along the length of the spool. The average pressure is identical on all lines such as XX in the diagram and as a consequence there is no net lateral force acting on the spool and the spool is therefore in equilibrium with no tendency to move in the lateral direction.

If now the spool is considered to be tapered, there is no longer a constant radial clearance throughout the annulus and the pressure distributions are different. The machining irregularities may be represented either as a spool which is of larger diameter at the end subject to high pressure $P_1$

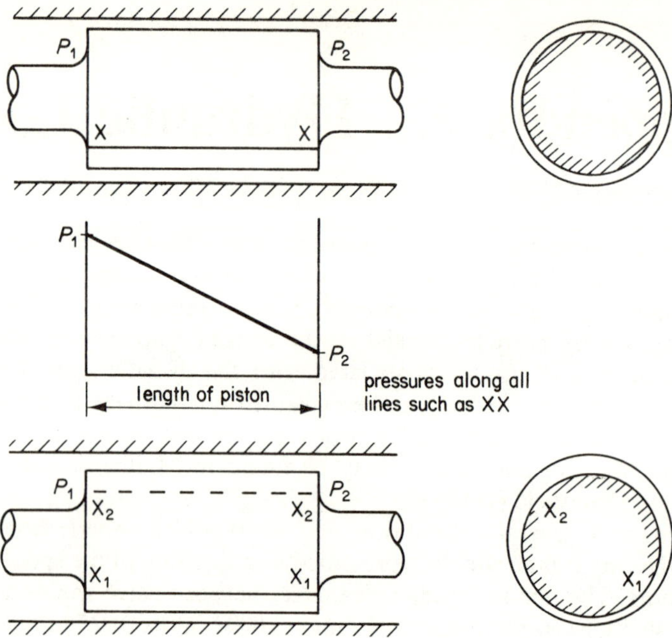



Figure E.1  Coaxial piston and bore

and smaller at the end subject to low pressure $P_2$ as shown in figure E.2 or alternatively as a spool which tapers the opposite way as shown in figure E.3. It is assumed that the centre line of each spool is parallel with the bore but that the spool is not located centrally. There is one point on the periphery where the distance between the spool and bore has its greatest value and a diametrically opposite point where this distance has its smallest value. These two positions are represented as A and B on figures E.2 and E.3. In each case, for the line AA′, the gap is approximately constant along the length of the spool and might vary from say 0.010 mm (0.0004 in) to say 0.013 mm (0.0005 in). However, the change along the opposite line BB′ is significant and the gap might vary from 0.001 mm (0.000 04 in) to 0.004 mm (0.000 16 in). With a constant gap (as line AA′) the pressure decreases along the length at a constant rate but if the gap varies then the rate at which the pressure decreases will also vary.

Figure E.2 illustrates a spool which tapers in the direction of the pressure. The pressure on line AA′ decreases at constant rate but the pressure on line BB′ decreases rapidly where the gap is small and then more slowly as the gap increases, giving a parabolic distribution of pressure illustrated in the lower diagram of figure E.2. Consequently the average pressure on line AA′ is greater than the average pressure on line BB′ and this implies

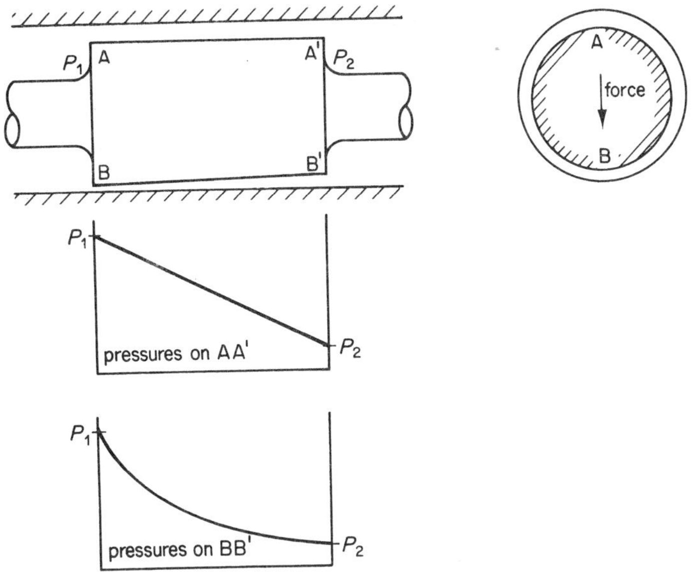


Figure E.2  Piston locking in bore

that there is a net force tending to move the spool laterally in the direction of B. As the spool is moved towards the cylinder wall, the gap along BB′ decreases. The pressure then decreases even more rapidly near the high-pressure end of the spool and the average pressure on BB′ decreases accordingly so that the lateral force further increases. The spool is therefore forced into close contact with the bore and a large axial force would be required to cause it to slide. This is 'hydraulic lock'.

Figure E.3 illustrates a spool which tapers the other way so that the gap decreases along the length of the spool. The rate at which the pressure decreases on line BB′ only becomes large near that end of the spool which is subjected to low pressure $P_2$. The average pressure on line BB′ is greater than the average pressure on line AA′ and the spool therefore tends to be centralised. This represents a case where the spool is self-centring and is the complete opposite from hydraulic lock.

Hydraulic lock is seen to occur if the geometry is such that a large pressure drop occurs at the end of the spool subject to high pressure over a small arc of the spool periphery. The cause may be irregularities on the surface of either the spool or the bore but there is also another common reason. If a layer of silt is deposited over some sector of the entrance to the annulus, then a low-pressure 'shadow' occurs downstream of the silt, giving pressure distributions over the affected arc similar to that shown in the lowest diagram of figure E.2.

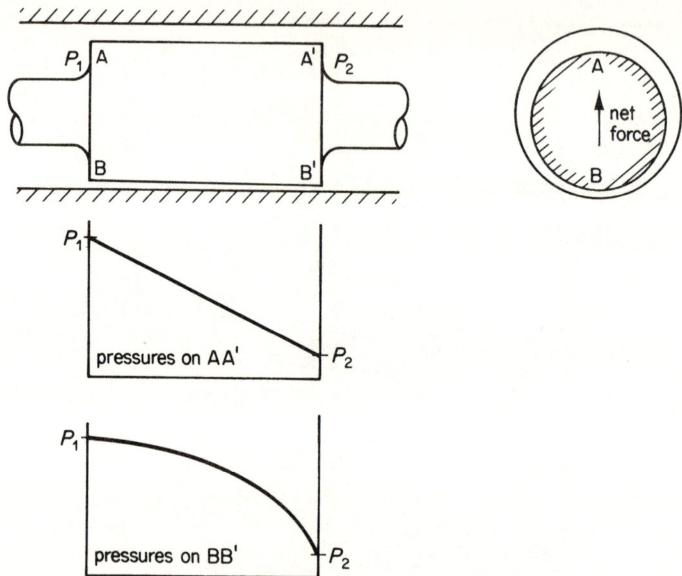



Figure E.3  Piston centralising in bore

The silt gives rise to the undesirable type of pressure distribution and the existence of particles in the fluid, even though they may be smaller in size than the nominal radial clearance, can be the cause of hydraulic lock.

To avoid hydraulic lock, it might be possible to make spools with favourable taper or even flexible cylindrical bores which bowed out under pressure. However, the usual precaution is to machine peripheral grooves in the spool at different places along its length. Each groove acts to equalise the pressure round the periphery in its own vicinity, thereby approximately equalising the average pressure acting on any line along the length of the spool.

# General Problems

1 An oil system is to be used for controlling the angular position of a platform whose mass is 2000 kg (4409 lb) and radius of gyration 0.6 m. A proposed design employs a variable pump fixed-displacement motor system with a mechanical linkage to vary the pump delivery in direct proportion to the difference between the angular positions of the platform and of an (input) handwheel. The pump is driven at constant speed and the motor shaft is directly connected to the platform. By assessing its characteristics analytically, investigate the suitability of a system having the parameters (i) pump flowrate $\pm\,4$ l/s (244 in$^3$/s) for every radian error, (ii) motor capacity 0.2 l/rad (12.2 in$^3$/rad), (iii) internal leakage total 0.005 l/s for every bar pressure difference between the two connecting pipelines (0.021 (in$^3$/s)/(lb/in$^2$)), (iv) trapped oil volume 13 l (793 in$^3$) and (v) viscous frictional drag on motor and platform total 22 (N m)/rad/s (195 (lb in)/(rad/s). For calculation purposes ignore force needed to move pump control and pressure losses in pipes.

(System unstable–need to reduce oil volume, incorporate gearing or decrease mechanical linkage ratio to stabilise. Note that an equation similar to 4.12 with coefficients

$$a_0 = \frac{V}{2\beta}\,\frac{\delta_m}{\alpha_p k}\,\frac{J}{\delta_m{}^2}\,, \qquad a_1 = \left(\frac{LJ}{\delta_m{}^2} + \frac{fV}{\delta_m{}^2\,2\beta}\right)\frac{\delta_m}{\alpha_p k}$$

$$a_2 = \left(1 + \frac{f}{\delta_m{}^2}\right)\frac{\delta_m}{\alpha_p k}$$

may be used with the Routh criterion.)

2 The piston rod of a double-acting hydraulic power cylinder is in the form of a rack which drives a rotary table through a pinion. Both sides of the cylinder are connected from a four-way valve through flexible hoses which dilate under pressure (the cylinder itself being rigid). The piston itself is 25 mm (0.984 in) long and is fitted in the cylinder without seals; the *radial* clearance between piston and cylinder is 0.125 mm (0.004 92 in). The piston diameter is 75 mm (2.95 in) and the rod diameter 25 mm (0.984 in). Estimate the frequency of the rotary oscillations of the table which would follow the sudden closing of the valve if the following data apply: maximum piston travel $\pm\,0.3$ m ($\pm\,11.8$ in); length of each flexible hose

1.8 m (70.9 in); internal hose diameter 25 mm (0.984 in); expansion of hose–volume increases by 1 part in 3500 for every bar change of pressure (1 : 50 750 for every lb/in$^2$); oil viscosity 25 cSt and specific gravity 0.85; table mass 90 kg (198.4 lb) and radius of gyration 127 mm (5 in); mass of piston and rack 50 kg (110.2 lb); number of teeth on table pinion 50; number of teeth on rack 2/cm (5.08/in); bulk modulus of oil 13.8 × 10$^8$ N/m$^2$ or 200 100 lb/in$^2$.

(About 13.7 Hz, the hydraulic frequency being about 14 Hz and the damping factor due to leakage alone about 0.2. Viscous drag is negligible. Effective mass with respect to piston 966.9 kg (2131 lb) and effective bulk modulus 5.13 × 10$^8$ N/m$^2$ (74 310 lb/in$^2$). Seal friction, etc., neglected.)

3 A control system to position a mass comprises an underlapped spool valve with full periphery ports and a double-acting piston with feedback lever as illustrated in figure 8.4. Estimate a suitable operating pressure for such a system to which the following data apply: diameter of spool valve 3 mm (0.118 in); amount of underlap $u$ 0.05 mm (0.001 97 in); net area of piston 0.002 5 m$^2$ (3.875 in$^2$); mass of load and piston 1000 kg (2205 lb); feedback link (see diagram) $\ell_1 = 5$ mm (0.197 in), $\ell_2 = 45$ mm (1.77 in); total oil volume 0.33 l (20.14 in$^3$); oil density 870 kg/m$^3$ (0.0314 lb/in$^3$); effective bulk modulus 14 × 10$^8$ N/m$^2$ (203 000 lb/in$^2$). Friction and leakage negligible.

(Less than 69 bar (1000 lb/in$^2$)—Routh stability criterion gives $K_c/K_q$ ratio indicating $P_s < 96$ bar (1400 lb/in$^2$)—damping factor approx. 0.1 at 69 bar and Bode diagram for this pressure gives about 3 dB gain margin.)

4 An electrohydraulic servo valve of the flow control type meters fluid to and from a double-acting power piston which is used to control the (rectilinear) position of a load. The position of the load is sensed by an electrical transducer and the control current on the servo valve coils is always directly proportional to the difference between an input voltage (representing the desired load position) and the voltage from the transducer. If the only damping obtained is that due to underlapping of the valve, estimate a suitable value of gain for the electrical amplifier which drives the valve coils (that is output voltage per volt error signal) if the following data apply: supply pressure 70 bar (1015 lb/in$^2$); main valve spool diameter 3 mm (0.118 in), symmetrical underlap at each port 0.05 mm (0.011 97 in), displacement 0.075 mm (0.002 95 in) for every milliampere coil current; valve coil impedance (treat as pure resistance) 500 Ω; transducer d.c. voltage output 0.04 V/mm (0.016 V/in); net area of power piston 0.0025 m$^2$ (3.875 in$^2$); mass of load and piston 1000 kg (2205 lb); trapped oil volume total 0.32 l (19.53 in$^3$);

effective bulk modulus $14 \times 10^8$ N/m$^2$ (203 000 lb/in$^2$); oil specific gravity 0.87.

(8 to give a gain margin of about 10 dB. Note that $K_q \approx 1.06$ (m$^3$/s)/m (1640 (in$^3$/s)/in) and $K_c \approx 3.77 \times 10^{-12}$ (m$^3$/s)/(N/m$^2$) (1.59 $\times 10^{-3}$ (in$^3$/s)/ (lb/in$^2$)); $\omega_h \approx 331$ rad/s; $\zeta \approx 0.1$; $K_{qe}/A \approx 2.5\,K_d$.)

5 Part of an oil hydraulic system consists of a large container fitted with a spring-loaded poppet valve as illustrated. There are varying flowrates into the container through the inlet and out from the container through the valve. The valve does not close completely and normal operation occurs with the valve open distance $x_0$ when inflow equals outflow $q_0$ and the pressure in the container is $P_0$. Making suitable simplifying assumptions, derive an (operator) relation between displacements $x$ of the poppet and changes in the entering flowrate $\overset{*}{q}$. Also assess whether stable operation can be expected if the following data apply: container volume $V$ 50 l (3051 in$^3$); $P_0$ 35 bar (507.5 lb/in$^2$); diameter $d$ of port 13 mm (0.512 in); effective mass $m$ of poppet 0.15 kg (0.331 lb); spring rate $k_s$ 500 N/cm (285.5 lb/in); precompression of spring $x_0$ 8.3 mm (0.327 in); viscous friction rate $f$ 9 N/(m/s) (0.0514 lb/(in/s); effective bulk modulus $\beta$ 13.8 $\times 10^8$ N/m$^2$ (200 100 lb/in$^2$); oil specific gravity 0.87.

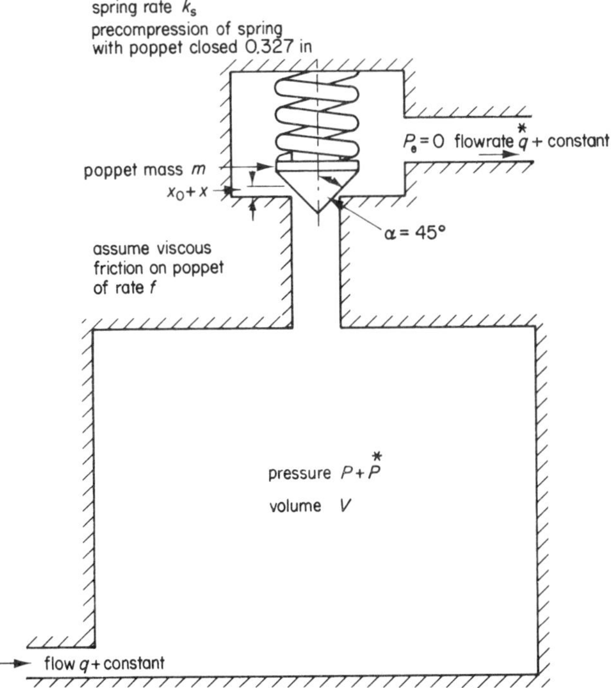

168

(We have

$$\frac{A'\beta}{V}q = \frac{A'\beta}{V}\left(\frac{q_0}{x_0} + \frac{1}{2}\frac{q_0}{P_0}\frac{k'}{A'}\right)x$$

$$+ \left(k' + \frac{1}{2}\frac{q_0}{P_0}\frac{f}{A'}\frac{A'\beta}{V}\right)\mathrm{D}x$$

$$+ \left(f + \frac{1}{2}\frac{q_0}{P_0}\frac{m}{A'}\frac{A'\beta}{V}\right)\mathrm{D}^2x$$

$$+ m\mathrm{D}^3x$$

where

$$A' = \frac{\pi d^2}{4} - \lambda x_0$$

For $\lambda$ see equation C.2

$$k' = k_\mathrm{s} + \lambda P_0$$

that is the governing equation is of the form

$$a_0\mathrm{D}^3 + a_1\mathrm{D}^2 + a_2\mathrm{D} + a_3$$

Numerically

$$a_1 \approx f, \qquad a_2 \approx k', \qquad a_3 \approx \frac{q_0}{x_0}\frac{A'\beta}{V}$$

and the system would be stable according to the criterion $a_1 a_2 > a_0 a_3$)

# References


Blackburn, J. F., Reethof, G., and Shearer, J. L. (1960). *Fluid Power Control*, MIT Technology Press, Cambridge, Mass.

Bowns, D. E., and Worton-Griffiths, J. (1972). The dynamic characteristics of a hydrostatic transmission system. *Proc. Inst. Mech. Engrs (London)*, **186**, No. 55, 755.

Brown, B. M. (1965). *The Mathematical Theory of Linear Systems*, Chapman and Hall, London.

Brown, F. T., and Nelson, S. E. (1965). Step responses of liquid lines with frequency-dependent effects of viscosity. *Trans. ASME (Am. Soc. Mech. Engrs), Ser. D: J. Basic Eng.*, 504.

Burrows, C. R. (1972). *Fluid Power Servomechanisms*, Van Nostrand, London.

Crandall, I. B. (1926). *Theory of Vibrating Systems and Sound*, Van Nostrand, London.

Dransfield, P. (1967–8). Hydraulic lock with single land pistons. *Proc. Inst. Mech. Engrs (London)*, **182**, part 1.

Foster, K. (1961). Dynamic analysis of a two-stage relief valve. *Proc. Conf. Oil Hydraulic Power Transmission*, Institution of Mechanical Engineers, London, p. 207.

Foster, K., and Parker, G. A. (1964–5). Transmission of power by sinusoidal wave motion through hydraulic oil in a uniform pipe. *Proc. Inst. Mech. Engrs (London)*, **179**, 599.

Healey, A. J., and Stringer, J. D. (1968–9). Dynamic characteristics of an oil hydraulic constant-speed drive. *Proc. Inst. Mech. Engrs (London)*, **183**, part 1.

Khaimovich (1965). *Hydraulic Control of Machine Tools*, Pergamon Press, Oxford.

Korn, J. (Ed.) (1969). *Hydrostatic Transmission Systems*, Intertext, London.

Leedham, V. E. (1965). *Pressure Waves in Spool Valve Drain Lines*, M. Eng. Thesis, University of Sheffield.

Lorenz, M. H., and Stringer, J. D. (1966). Pressure waves in a pipeline closed at one end. *The Engineer*, June 24th, 965.

———— (1966–7). Oil hydraulic spool valve operating times. *Proc. Inst. Mech. Engrs (London)*, **181**, part 1, No. 4, 75.

McCloy, D., and McGuigan, R. H. (1965). Some static and dynamic characteristics of poppet valves. *Proc. Conf. Advan. in Auto. Control*, Institution of Mechanical Engineers, London, paper 23, p. 79.

McCloy, D., and Martin, H. R. (1973). *The Control of Fluid Power*, Longman, London.

Manhajm, J., and Sweeney, D. C. (1955). An investigation of hydraulic lock. *Proc. Inst. Mech. Engrs (London)*, **169**, No. 42, 865.

Martin, K. F. (1970). Stability and step response of a hydraulic servo with special reference to unsymmetrical oil volume conditions. *J. Mech. Eng. Sci.*, **12**, No. 5, 331.

Merritt, H. E. (1967). *Hydraulic Control Systems*, Wiley, New York.

Morse, A. C. (1963). *Electrohydraulic Servos*, McGraw-Hill, New York.

Nikiforuk, P. N., Ukrainetz, P. R., and Tsai, S. C. (1969). Detailed analysis of a two-stage four-way electrohydraulic flow control valve. *J. Mech. Eng. Sci.*, **11**, No. 2, 168.

Royle, J. K. (1959). Inherent non-linear effects in hydraulic control systems with inertia loading. *Proc. Inst. Mech. Engrs (London)*, **173**, No. 9, 257.

———— (1961). Hydraulic damping techniques at low velocity. *Proc. Conf. Oil Hydraulic Power Transmission*, Institution of Mechanical Engineers, London, paper 10.

Schlösser, W. M. J. (1969). The overall efficienty of positive-displacement pumps. *BHRA Fluid Power Symp.*, paper SP 983.

Shearer, J. L. (1954). Dynamic characteristics of valve-controlled servo motors. *Trans. ASME (Am. Soc. Mech. Engrs)*, **76**, 895.

Shell International Petroleum Co. Ltd (1967). *Technical Data on Shell Tellus Oils*, 2nd edn, London.

Shute, N. A., and Turnbull, D. E. (1961). Some trends in electrohydraulic servo valve design. *Proc. Oil Hydraulics Conf.*, Institution of Mechanical Engineers, London, paper 11, p. 130.

Smith, L. H., *et al.* (1960). Hydraulic fluid bulk modulus. *Natl Conf. on Industrial Hydraulics*.

Takenaka, T., and Urata, E. (1968). Static and dynamic characteristics of oil hydraulic control valves. *Proc. Fluid. Power Int. Conf.*, 1968, p. 67.

Walters, R. (1967). *Hydraulic and Electrohydraulic Servo Systems*, Iliffe, London.

West, J. C. (1953). *A Textbook of Servomechanisms*, English Universities Press, London.

Zeller, J. R. (1968). Design and analysis of a modular servo amplifier for fast-response electrohydraulic control systems. *NASA Tech. Note*, No. TN D–4898.

# Index